MOLLUSCS: CAUDOFOVEATA, SOLENOGASTRES, POLYPLACOPHORA AND SCAPHOPODA

[redacted]

A NEW SERIES

Synopses of the British Fauna

Edited by Doris M. Kermack AND R. S. K. Barnes

The *Synopses of the British Fauna* are illustrated field and laboratory pocket-books designed to meet the needs of amateur and professional naturalists from sixth-form level upwards. Each volume presents a more detailed account of a group of animals than is found in most field-guides and bridges the gap between the popular guide and more specialist monographs and treatises. Technical terms are kept to a minimum and the books are therefore intelligible to readers with no previous knowledge of the group concerned.

Volumes 1–18 inclusive are available from the Linnean Society of London, Burlington House, Piccadilly, London W1V 0LQ.
Volumes 19–28 inclusive are available from the Cambridge University Press.
All subsequent volumes may be obtained from E. J. Brill, Publishing Company, Leiden, The Netherlands via Natural History Book Service, 62 Tritton Road, London SE21 8DE.

1. *British Ascidians* R. H. Millar
2. *British Prosobranchs* Alastair Graham
3. *British Marine Isopods* E. Naylor
4. *British Harvestmen* J. H. P. Sankey and T. H. Savory
5. *British Sea Spiders* P. E. King
6. *British Land Snails* R. A. D. Cameron and Margaret Redfern
7. *British Cumaceans* N. S. Jones
8. *British Opisthobranch Molluscs* T. E. Thompson and Gregory H. Brown
9. *British Tardigrades* C. I. Morgan and P. E. King
10. *British Anascan Bryozoans* J. S. Ryland and P. J. Hayward
11. *British Freshwater Bivalve Mollusca* A. E. Ellis
12. *British Sipunculans* P. E. Gibbs
13. *British and Other Phoronids* C. C. Emig
14. *British Ascophoran Bryozoans* P. J. Hayward and J. S. Ryland
15. *British Coastal Shrimps and Prawns* G. Smaldon
16. *British Nearshore Foraminiferids* John W. Murray
17. *British Brachiopods* C. H. C. Brunton and G. B. Curry
18. *British Anthozoans* R. L. Manuel
19. *British Planarians* I. R. Ball and T. B. Reynoldson
20. *British Pelagic Tunicates* J. H. Fraser
21. *British and Other Marine and Estuarine Oligochaetes* R. O. Brinkhurst
22. *British and Other Freshwater Ciliated Protozoa: Part I* C. R. Curds
23. *British and Other Freshwater Ciliated Protozoa: Part II* C. R. Curds, M. A. Gates and D. McL. Roberts
24. *British Nemerteans* R. Gibson
25. *Shallow-water Crabs* R. W. Ingle
26. *Polyclad Turbellarians* S. Prudhoe
27. *Tanaids* D. M. Holdich and J. A. Jones
28. *Free-living Marine Nematodes: Part I British Enoplids* H. M. Platt and R. M. Warwick
29. *Siphonophores and Velellids* P. A. Kirkpatrick and P. R. Pugh
30. *Euphausiid, Stomatopod and Leptostracan Crustaceans* J. Mauchline
31. *Earthworms* R. W. Sims and B. M. Gerard
32. *Polychaetes: British Amphinomida, Spintherida & Eunicida* J. D. George and G. Hartmann-Schröder
33. *Ctenostome Bryozoans* P. J. Hayward
34. *Cyclostome Bryozoans* P. J. Hayward and J. S. Ryland
35. *Millipedes* J. Gordon Blower
36. *Halacarid Mites* J. Green and Miranda Macquitty

Synopses of the British Fauna
Edited by Doris M. Kermack and R. S. K. Barnes
No. 37

MOLLUSCS: CAUDOFOVEATA, SOLENOGASTRES, POLYPLACOPHORA AND SCAPHOPODA

Keys and notes for the identification of species

A. M. JONES and J. M. BAXTER
Department of Biological Sciences, The University, Dundee, DD1 4HN, Scotland

TUTA SUB AEGIDE PALLAS · EJB ·

1987
Published for
The Linnean Society of London
and
The Estuarine and Brackish-water Sciences Association
by
E. J. Brill/Dr W. Backhuys
London – Leiden – Köln – København

ISBN 90 04 08197 6

Printed in Great Britain at The Bath Press, Avon

Molluscs: Caudofoveata, Solenogastres, Polyplacophora and Scaphopoda

A. M. JONES and J. M. BAXTER
Department of Biological Sciences, The University, Dundee, DD1 4HN, Scotland

Contents

Foreword vii

General Introduction 1

CAUDOFOVEATA
- Introduction 2
- General Structure 3
- Biology and Ecology 6
- Collection and Preservation 7
- Classification 8
- Key 9
- Systematic part 10

SOLENOGASTRES
- Introduction 18
- General Structure 19
- Biology and Ecology 22
- Collection and Preservation 23
- Classification 24
- Key 25
- Systematic part 26

POLYPLACOPHORA
- Introduction 38
- General Structure 39
- Biology and Ecology 42
- Collection and Preservation 44
- Classification 45
- Key 47
- Systematic part 49

SCAPHOPODA
- Introduction 91
- General Structure 92
- Biology and Ecology 94
- Collection and Preservation 95
- Classification 96

Key 97
Systematic part 98

Glossary of special terms 108

Acknowledgements 111

References cited 112

APPENDIX Distribution maps 114

Index 122

Plates 1–3 between pp. 56–57

Foreword

The three best-known mollusc classes, the Gastropoda, Bivalvia and Cephalopoda, contain between them 98.5% of all living mollusc species. It is less generally appreciated, however, that there are five other classes of the Mollusca, of which four are represented around the coast of Britain. This *Synopsis* covers these four groups, which, although not particularly closely related, are each so small in numbers of contained species as to make it uneconomic to treat them separately in this series.

Two of them, the Caudofoveata and Solenogastres, appear at first sight to be most unlike molluscs, in that they are vermiform animals lacking shells, a foot and a distinct head. Some authorities consider them to be survivors from the evolutionary phase before the evolution of the characteristic molluscan shell from a series of separate scales. The outer coverings of the other two groups are also distinctive. The polyplacophorans, as their name suggests, possess multiple shells in the form of eight separate dorsal plates; whilst the single shell of the scaphopods forms an elongate and often curved tube, open at each end.

For taking on all four groups, we are indebted to Drs Jones and Baxter. Their treatment not only will permit the relatively familiar chitons and the other much less familiar classes to be identified, but also will, we hope, introduce the members of these groups to a much wider audience, including marine biologists. The authors have, in their text and illustrations, shown these interesting but neglected animals to be well worthy of such a wider appreciation.

R. S. K. Barnes
Estuarine & Brackish-water
Sciences Association

Doris M. Kermack
The Linnean Society
of London

General introduction

The classes considered in this *Synopsis* are often referred to as the minor Molluscs being, for the most part, relatively limited in diversity and ecological significance. While the class status of the Polyplacophora and the Scaphopoda has been well established, the status of the Caudofoveata and the Solenogastres remains under debate. These two groups were placed together in the class Aplacophora until Salvini-Plawen (1969) divided the Aplacophora into two independent classes, the Caudofoveata (=Chaetodermamorpha) and the Solenogastres (=Neomeniamorpha). Recent works by Scheltema (1978) and Ivanov (1981) have argued for a restoration of the class Aplacophora with the Caudofoveata and the Solenogastres being given sub-class status. We have retained Salvini-Plawens' classification for this volume as there is no definitive opinion in this matter but we do not wish this to be taken to imply that this is necessarily more correct. Three of the classes considered are confined to sub-littoral habitats and in many cases they extend into deep oceanic environments. We have adopted an arbitrary depth limit of 250 m to define the lower limit of the zone from which species were selected for inclusion in this *Synopsis*.

CAUDOFOVEATA

Introduction

The Caudofoveata (formerly Aplacophora; Chaetodermamorpha) comprise a small class of exclusively burrowing, worm-shaped molluscs. They are regarded as representing the earliest offshoot within the phylum and Salvini-Plawen (1975) regards them as being quite distinct phylogenetically from the remaining molluscan classes. The burrowing habit of this group has resulted in only the cerebrally-innervated portion of the archimolluscan foot being retained as the foot shield; it has also resulted in more general modifications so that their molluscan character was in doubt until the work of Hoffman (1949).

General structure

The external appearance of a caudofoveate is well illustrated by *Chaetoderma nitidulum* (Fig. 1A), the commonest British species. They are wormlike in appearance with a bilaterally symmetrical cylindrical body up to about 80 mm in length, of which the subrostral **foot-shield** (Fig. 1C) is the only remaining rudiment of the foot. The **pallial complex** or **dorsal sense organ** is posterior and terminal in position (Fig. 1B). The body lacks a definite head or other sub-division and often the anterior and posterior ends are not distinguishable on casual examination. In *C. nitidulum*, the anterior end is marked by a constriction (Fig. 1A) but this does not morphologically distinguish a true head region. The mouth and anus are terminal at opposite ends of the body. The body may be divided into 5 approximate regions (Fig. 1D) for describing the distribution of external features. The body is bounded externally by a chitinous **cuticle** and is covered with calcareous spicules which occur as **scales** and, at the posterior end, as relatively large **spines**: the shape and distribution of these structures is species-specific. The foot-shield (Fig. 1C) and the pallial cavity (Fig. 1F) are the only regions left uncovered by the scales/spines. The foot-shield (Fig. 1C) has a cuticularised ciliar epithelium and typically lies in a post-oral position with antero-lateral extensions which flank the mouth: the shape and position of the foot-shield is again characteristic for each species although the process of fixation and its consequent muscular contraction may distort the shape considerably. The foot-shield contains numerous sensory cells and mucous cells and appears to be a glandulosensory area associated with food selection and intake.

The posterior, bell-shaped **pallial cavity** (Fig. 1F) is terminal in position and always contains one pair of lamellate gills (**ctenidia**) which project to the exterior and in which the outer lamellae are larger than the inner ones. The pallial cavity also contains a medio-ventral **anus** and two mucous ducts which open ventrolaterally; the latter form the paired secondary **gonoducts**.

The alimentary tract begins with an evertible **buccal cavity** which leads into a large **pharynx** provided with glandular organs. There is a ventral **radula** in which the teeth (Fig. 1G) have, in some species, been reduced to small denticles only (Fig. 1H, I); this is correlated with the enlargement of the outer supports and the evolution of the median element into a very large **basal plate**. Only the radular teeth are normally sclerotised although the tips of the accessory structures may be reinforced also. The basal plate normally lies in a plane at right angles to that of the radular teeth but, after maceration or enclosing between two slides, it becomes orientated into the same plane; all diagrams present the arrangement as seen in this secondary position, therefore. The postradular **oesophagus** leads into a short, broad **midgut** which divides into the large, medioventral **digestive gland** and thus to the uncoiled, narrow **intestine** and the terminal **rectum**.

Caudofoveates are **dioecious**, the long dorsal **gonads** usually being fused into a single structure. Gametes are shed into the **pericardium** and, via **pericardioducts**, to the glandular mucous ducts from where they are released directly to the sea.

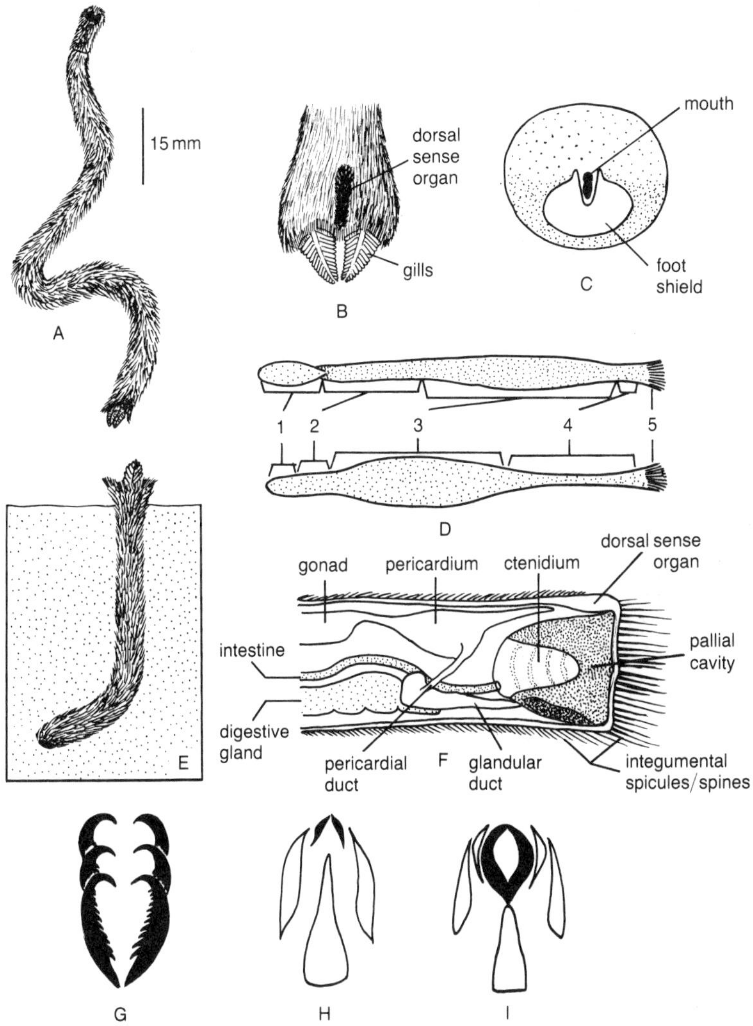

Fig. 1. General features of the Caudofoveata. A. External appearance of *Chaetoderma nitidulum*. B. Enlarged view of the posterior end of *C. nitidulum* to show the bi-pectinate ctenidia. C. Anterior end of *C. nitidulum* to show foot shield and mouth. D. Partition of the caudofoveate body into regions: 1, foregut region; 2, midgut region; 3, digestive gland region; 4, prepallial region; 5, pallial region. E. Position of living animal in the sediment. F. Organisation of posterior body and pallial complex. G, H and I. Radula apparatus of *Scutopus*, *Chaetoderma* and *Falcidens* respectively to show differing degrees of tooth (black) development and development of supporting structures. (Figs. D–I after Salvini-Plawen 1975).

Biology and ecology

The Caudofoveata are exclusively marine animals, geographically limited by a minimum salinity requirement of 28–30‰. They appear to be world-wide in distribution but the group is best known from European seas. They are found burrowed in muddy sediments in water more than 20 m in depth; their abundance appears to be very dependent on local conditions although Salvini-Plawen (1975) considers that they '... can probably be found on nearly every kind of muddy bottom without regard to geographical region ...': the limited distribution records are considered to be attributable largely to the restricted collection methods available. An average abundance of 1 per square metre in much of the North Sea is recorded for *C. nitidulum*, but under favourable conditions caudofoveates may reach densities of 4 or 5 per square metre.

Because of their habitat and life style, these animals are usually known only from preserved specimens obtained by dredging and little is known of their physiology and behaviour. They live in burrows constructed by driving their anterior end into the sediment by muscular contractions and expansions. They live in a vertical burrow (Fig. 1E) with the anterior end down and the expanded gills at its entrance; when disturbed, the gills are retracted and the animal retreats further into its burrow. Caudofoveates are very sluggish and often remain immobile for long periods but they can creep slowly using the anterior portion of the body.

Reproduction of these dioecious animals is by external fertilisation but the larval development has not been described. Nothing is known on growth rate or longevity.

They appear to feed on micro-organisms and organic matter selected from their surroundings but they do not appear to ingest unsorted sediment. The radula acts to comminute food particles selected by the glandulosensory foot-shield.

The burrowing habit of caudofoveates indicates perception of gravity and possibly light but little or no evidence exists on the sensory responses of these animals. There are no eyes, tentacles or statocysts.

Collection and preservation

Collection must be by dredging or possibly the use of grab sampling. Sediment collected must be carefully sorted, preferably under a low-power microscope. Whenever possible specimens should be examined alive but if fixation is carried out, it must be in a fixative which will not destroy the calcareous scales and spines required for identification. The best fixatives, therefore, are 4% neutral formaldehyde in seawater or 70% alcohol. To examine the radula, the best procedures are either (a) maceration of the animal in 10% hot potassium hydroxide or (b) compression between two slides after clearing the whole animal in methyl benzoate or Cedarwood oil. The foot-shield is best examined in live specimens and it should be noted that its shape does change somewhat after fixation. Examination invariably requires the use of at least a good quality dissecting microscope.

Classification

The taxonomy of this group is still fluid. The classification given below is according to Salvini-Plawen but Scheltema (1978) and Ivanov (1981) have proposed that both the Caudofoveata and the Solenogastres should be given sub-class status only within the class Aplacophora. Ivanov (1981) has also indicated that the genus *Chaetoderma* is in need of revision.

Class CAUDOFOVEATA

Family Limifossoridae Salvini-Plawen 1968
Scutopus ventrolineatus Salvini-Plawen 1968

Family Chaetodermatidae Ihering 1876
Chaetoderma nitidulum Lovén 1845
Falcidens crossotus Salvini-Plawen 1968

Key

1. Ventral furrow absent. A pair of bi-pectinate gills. Mouth surrounded by foot-shield and with a radula present. Mouth and anal chamber terminal in position. Spicula squamate, tapering, spearshaped and half-raised ........................CAUDOFOVEATA **2**

Ventral furrow present. Gills, when present, developed as folds, not bi-pectinate. Mouth without foot-shield. Mouth and anal chamber may be terminal and a radula may be present. ....SOLENOGASTRES (p.18)

2. Body clearly divided into regions of different diameter and with a posterior region bearing a terminal orange 'tassel' ....*Falcidens crossotus* (p. 16)

Body of uniform diameter throughout **3**

3. Body slender, up to 80 mm in length, with a definite transverse constriction between the first and second body regions. Pallial cavity bell-shaped; foot-shield round in shape with the mouth in a deep frontal cleft; radula reduced to a single pair of denticles ....*Chaetoderma nitidulum* (p. 14)

Body slender, up to 35 mm in length, with a distinct mid-ventral line or suture in the anterior part of the middle region. Foot-shield frontal in position with the mouth at its posterior margin; radula distichous with 10–16 transverse rows of large, well-developed teeth ....*Scutopus ventrolineatus* (p. 10)

Family LIMIFOSSORIDAE Salvini-Plawen, 1968

Animals with a short, sometimes blunt, body and often with a ventral suture. The foot-shield is homogeneous, post-oral in position and may be divided or paired. The radula is large, bipartite and bears several transverse rows of often denticulated teeth; there are no additional cuticular supports. The digestive midgut gland is well-developed.

Genus SCUTOPUS Salvini-Plawen, 1968

Uniformly shaped, cylindrical animals with a homogeneous foot shield. Radular teeth with median denticles but without medial wings at base. Digestive midgut gland long.

Type species: *Scutopus ventrolineatus* Salvini-Plawen.

Scutopus ventrolineatus Salvini-Plawen, 1968
(Fig. 2)

A slender animal with undifferentiated body regions. This species has the ability to coil up, a feature unusual for a caudofoveate. The anterior portion of the middle body region bears a distinct mid-ventral line or suture (Fig. 2A) which has its own associated scale-types (Figs. 2G, e and h). Animals range up to 35 mm in length but contraction and extension on fixation may cause halving or doubling of the length respectively. The length : breadth ratio is 20–35.

The foot-shield (Fig. 2B–E) is confined to a post-oral region but it assumes a dorso-frontal position (Fig. 2B) so that in living specimens it appears in a frontal position. The contraction resulting from fixation may cause a change in shape (Fig. 2C and D) so that it comes to resemble a divided plate. The distichous radula bears 10–16 transverse rows of large, well-developed teeth (Fig. 2F), each ending in an inwardly-curved hook and bearing 9–11 curved denticles on the inner edge. These teeth are only partially sclerotised and only the frontal 2–4 pairs of teeth having definite strengthening; this feature makes examination of the radula quite difficult.

The pallial region is undifferentiated with no particular colour. The ctenidia are small, medially bent and have 6–8 distinct lamellae only at the outer edges.

The body is covered by almost clinging scales (Fig. 2G) which are more slenderly formed and lanceolate in shape in anterior regions (Fig. 2G a–k) while the posterior ones have a basally rounded, tapering form (Fig. 2G l–o).

Small specimens are usually a whitish or brownish colour while larger animals are more distinctly brown, at least in the digestive gland region. The

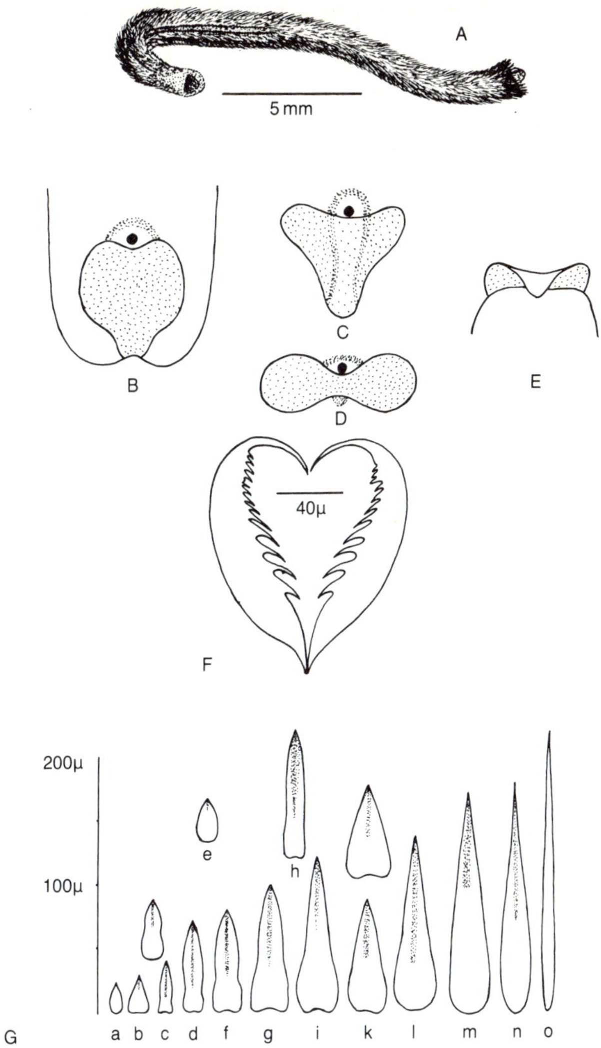

Fig. 2. *Scutopus ventrolineatus*. A. External appearance of adult specimen. B. Diagrammatic view of anterior end of living specimen seen from dorsal side. C and D. Different shapes assumed by the foot shield in the contracted condition. E. Ventral view of anterior end of living specimen. F. One transverse row of radula teeth. G. The range of scale types: a–c, foregut region; d–k, midgut and digestive gland regions; l–o, prepallial and pallial regions; e and h are from the ventral groove. (A after Salvini-Plawen 1971; B–G after Salvini-Plawen 1975).

short pre-pallial region is markedly lighter in colour while a pinkish tinge is often associated with the ctenidia and the anterior region of the body.

S. ventrolineatus is found in soft, muddy bottoms at depths greater than 40 m where it is regularly distributed with very high population densities locally. It is a typical East-Atlantic species recorded from Loch Linnhe (110–115 m), Loch Creran (40 m) and Firth of Lorne (50 m) on the west coast of Scotland and the Irish Sea just off the Isle of Man (70 m).

Family CHAETODERMATIDAE Ihering, 1876

Body either uniformly cylindrical or sometimes with a tail-like posterior. The foot-shield is undivided and surrounds the mouth. The radula is much reduced and consists of only one transverse row of real teeth if not completely reduced and lacking. One or two pairs of cuticular lateral supports and a large basal plate; some further, less obvious, cuticular strengthening may be present. The digestive midgut gland is well developed. Two genera are recognised by Salvini-Plawen but Ivanov (1981) has questioned the validity of *Falcidens*.

Genus CHAETODERMA Lovén, 1845

Uniformly cylindrical body, sometimes with an anteriorly placed constriction. The radular teeth are represented only by pairs of isolated denticles without basal connections or are entirely lacking; lateral supports simple when present, basal plate strongly developed. The foot-shield is homogeneous and the digestive midgut gland is long. This genus is a heterogeneous group requiring revision (Ivanov, 1981).

Type species: *Chaetoderma nitidulum* Lovén, 1845.

Chaetoderma nitidulum Lovén, 1845
(Fig. 3)

Crystallophrisson nitens Möbius, 1875
Crystallophrisson nitidulum Winckworth, 1951

A large (up to 80 mm long) caudofoveate with a homogeneous body shape but sometimes with a definite constriction towards the anterior end and a slightly swollen mid-region (Fig. 3A). The length : breadth ratio is 15–25. The pallial cavity is generally enlarged with distinct orange-brown secretions. The number of each type of scale within a body region depends largely on size and age. All scales and spicules (Fig. 3G) have a definite keel except for the small squamiform scales (Fig. 3G a) and bomb-like (Fig. 3G b) spicules. The foot-shield (Figs. 3B–D) is generally round in shape with the mouth in a deep frontal cleft. The digestive midgut is long; the ctenidia are large and have up to 25 alternating lamellae on each side.

The radula apparatus (Figs. 3E and F) carries one pair of denticles and the basal plate is heavily strengthened by transverse ridges. The two lateral supports have distal strengthenings.

The body is brownish-grey veiled by a silver lustre originating from the scales. The anterior region may show a slight pinkish tinge. The ctenidia are red in colour and the bell-shaped pallial region is covered by orange-brown secretions.

C. nitidulum is distributed in sediments varying from muds to sands in depths of 8–910 m but is mainly found in shallow open sea areas 20–150 m in depth. It is widely distributed with densities of about 1 per square metre.

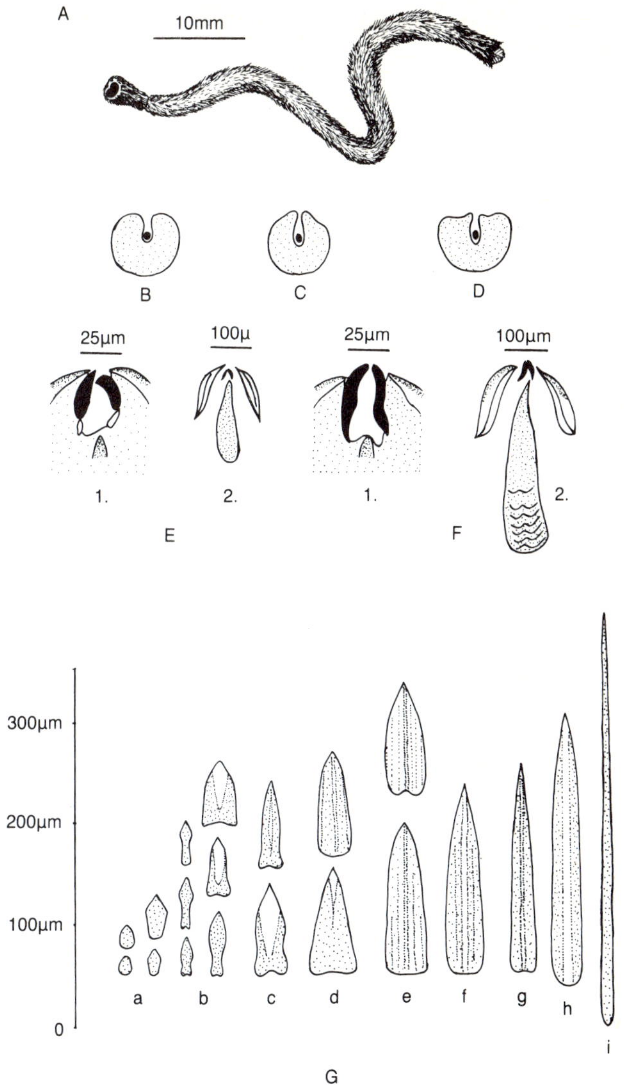

Fig. 3. *Chaetoderma nitidulum* A. External appearance of adult specimen. B–D. Foot shield in three different conditions of contraction. E. Radula of juvenile specimen: 1, detail of the radula tooth (black); 2, overall view. F. Radula of adult specimen: 1, detail of the radula tooth (black); 2, overall view. G. Range of scale types: a, foregut region; b–c, midgut region; d–g, digestive gland region; h–i, prepallial and pallial regions. (A after Salvini-Plawen 1971, B–G after Salvini-Plawen 1975).

Genus FALCIDENS Salvini-Plawen, 1968

Body either uniformly cylindrical or with a distinctly narrower, tail-like posterior region. Always with two, sickle-like radula teeth placed on tip of the basal plate, sometimes by means of a proximal symphyse. The foot shield is homogeneous and may entirely surround the mouth. The digestive midgut gland may be short or long.

Type species: *Falcidens crossotus* Salvini-Plawen 1968.

Falcidens crossotus Salvini-Plawen, 1968
(Fig. 4)

An unmistakable caudofoveate with a characteristic body shape (Fig. 4A), the posterior region being slender and tail-like and terminating in an orange coloured pallial 'tassel': this posterior region may be coiled. This body-form is unique among British caudofoveates. Extended animals range in length up to 25 mm but are commonly only 15–20 mm in length: the degree of contraction due to fixation is very important and fully contracted animals may be only half of the length of fully extended ones.

The foot-shield (Fig. 4B–D) surrounds the mouth posteriorally and laterally but does not fuse pre-orally. The radula comprises only a single pair of sickle-shaped radular teeth (Fig. 4E) which are pointed at their bases and which do not have a symphyse. The lateral supports are simple and, like the basal plate, are only strengthened distally.

The ctenidia are large with up to 11 to 12 alternating lamellae. The digestive gland is short, not entering the slender posterior region. The middle of the body is covered with characteristic, clinging scales (Fig. 4F, f–h).

The body is brownish or blackish veiled by a distinct silver lustre imparted by the spicules. The slender posterior region is normally lighter in colour and ends in an orange pallial 'tassel'. The ctenidia are pink in colour.

F. crossotus is found in soft, muddy bottoms at depths of between 30 and 470 m, with a maximum density of up to 3 per square metre. The only record from around the British coast is that from 50 m depth in the Firth of Lorne on the west coast of Scotland (Gage, 1972).

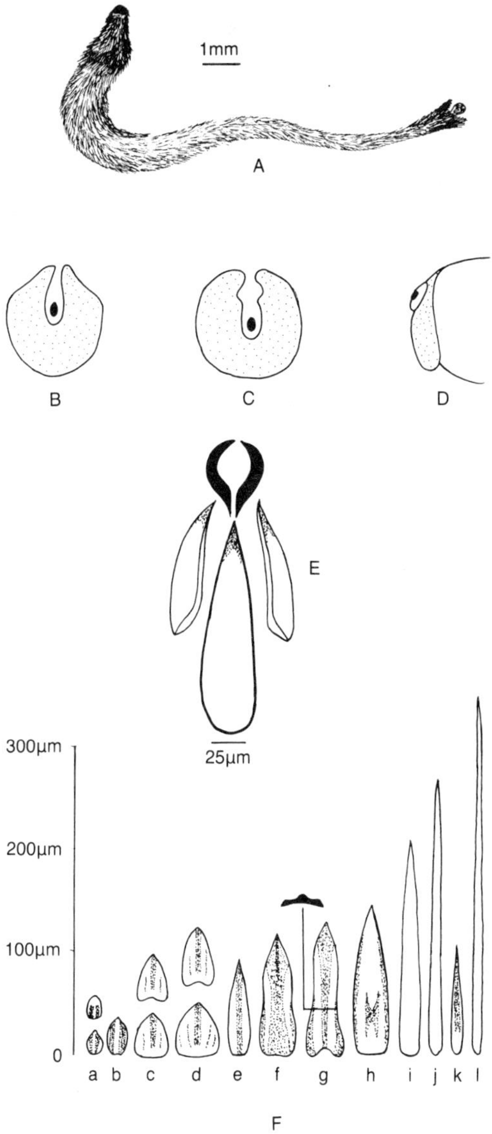

Fig. 4. *Falcidens crossotus*. A. External appearance of adult specimen. B and C. Frontal view of foot shield in different states of contraction. D. Lateral view of foot shield and buccal area. E. Radula (teeth in black). F. Range of scale types: a–b, foregut region; c–d, midgut region; d–f, digestive gland region; g–l, in the tail-like prepallial and pallial regions. (A after Salvini-Plawen 1971, B–F after Salvini–Plawen 1975).

SOLENOGASTRES

Introduction

The Solenogastres (formerly Aplacophora; Neomenioidea) are a small class of shell-less marine molluscs which live on the surface of mud or on cnidarians which form their principal food source. They are regarded by Salvini-Plawen (1975) as being more closely related to the shell-bearing molluscs than to the Caudofoveata with which they were once grouped. The systematics of this group have been revised by Salvini-Plawen (1978).

General structure

The Solenogastres are generally small (less than 50 mm long) animals with a bilaterally symmetrical body which may be either long and slender or short and relatively broad. The body (Fig. 5A) lacks a definite head or other subdivision and often careful examination is required to distinguish anterior and posterior ends. A mid-dorsal longitudinal keel or crest is often present (Fig. 5B). The ventral surface (Fig. 5A) has a median longitudinal groove, the **pedal groove**, which terminates anteriorally in a concavity, the **pedal pit**, and posterially runs into the **anus**: this groove contains a central ridge which is considered to represent the molluscan foot. The **mouth** is generally a subterminal, antero-ventral vertical slit which leads into a large cavity, the **vestibule** or atrium. The anus opens into a terminal posterior cavity which is a **cloaca**. Bi-pectinate **ctenidia** are never present in Solenogastres but, in the Neomeniidae, the cloaca contains secondary respiratory structures composed of lamellae, plicae or papillae.

The dorsal, lateral and greater part of the ventral surfaces are covered by a **cuticle** with a complex structure (Fig. 5C) and, in some groups, epidermal papillae are found; these are long stalks of cellular epidermis which ascend from the epidermis through the cuticle to terminate at the surface in a spherical, vacuolated vesicle (Fig. 5D). The cuticle in all British Solenogastres bears one or several layers of calcareous **spicula** which are mainly very small and require microscopic examination. They may be set vertically (radial) but are usually backward-pointing (tangential) and make only a small angle with the cuticle surface. The spicula may be hollow (open) or solid in construction and many are channel-shaped. In some species, special spicules may be found associated with the cloaca and are presumed to be copulatory in function.

The pedal groove and pedal pit (Fig. 5A and E) are devoid of spicules and epidermal papillae but the pedal groove may have accessory folds on each side of the pedal groove proper as in *Neomenia carinata*. The pedal pit is lined by a heavily ciliated epidermis and both it and the pedal groove are supplied with numerous secretory glands which produce mucus. *Eleutheromenia* and *Lepidomenia* both have a radula but the other British species lack this structure; both *Eleutheromenia* and *Lepidomenia* have pairs of teeth in transverse rows, a condition known as **distichous**, but it does tend to be rather variable.

The Solenogastres are **hermaphrodite** with a pair of dorsal hermaphrodite glands, the **gonads** discharging into a ciliated **gonopericardial canal** leading to the **pericardial cavity**: from here the paired **gonoducts** pass first anteriorally and then posteriorally, bearing vesicles presumed to be **seminal vesicles** and **seminal receptacles** (Fig. 5E). They then pass through glandular **shell glands** after which the paired gonoducts fuse to open into the cloaca just below the anus; a copulatory armature may be present. The circulatory system is open with a **heart** containing a median ventricle and a pair of auricles, the

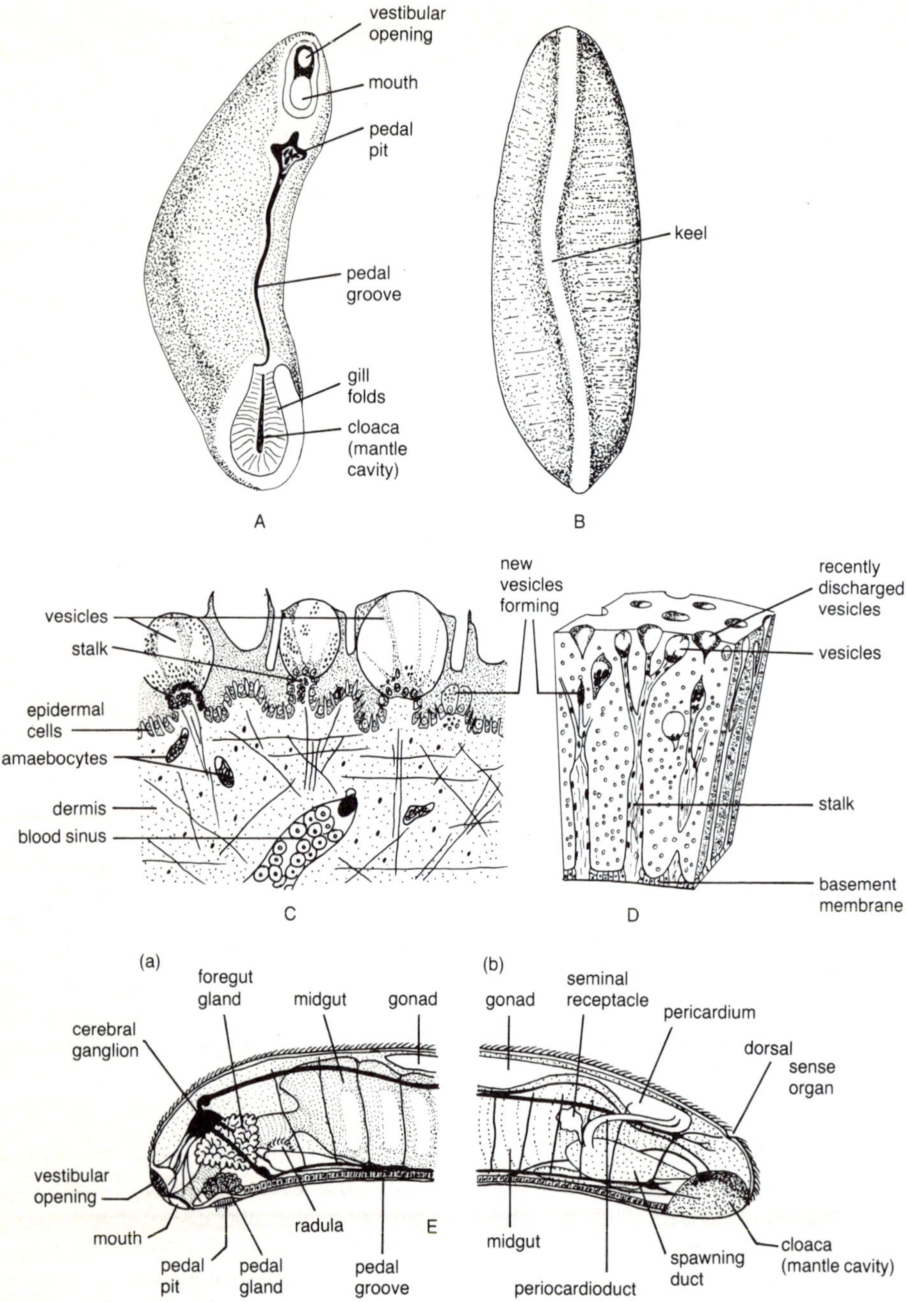
vestibular opening
mouth
pedal pit
pedal groove
gill folds
cloaca (mantle cavity)
A
keel
B
vesicles
stalk
epidermal cells
amaebocytes
dermis
blood sinus
new vesicles forming
C
recently discharged vesicles
vesicles
stalk
basement membrane
D
(a)
foregut gland
midgut
gonad
cerebral ganglion
vestibular opening
mouth
pedal pit
pedal gland
radula
pedal groove
E
(b)
gonad
seminal receptacle
pericardium
dorsal sense organ
midgut
periocardioduct
spawning duct
cloaca (mantle cavity)

whole being largely fused with the pericardium. True excretory organs are absent. For a more detailed account of the internal anatomy, see Hyman (1967).

Fig. 5. Organization of Solenogastres. A. Ventral view of *Neomenia carinata* to show the pedal groove and associated structures. B. Dorsal view of *N. carinata* to show the dorsal keel. C. Epidermal papillae in *Proneomenia antarctica*. D. Section through the body wall of *N. carinata*. E. Organization of (a) anterior and (b) posterior regions of a generalized Solenogastre. (A and B after Hansen 1889; C and D after Hoffman 1949; E after Salvini-Plawen 1971).

Biology and Ecology

The Solenogastres are an entirely marine benthic group found from shallow water to considerable depths; they are rarely found in water less than 20 m deep. *Lepidomenia hystrix* has been recorded from the littoral zone in Strangford Lough, Northern Ireland, the only littoral record for a Solenogastre.

They do not normally burrow and are found entwined about other organisms, usually hydroids, corals, alcyonarians and zooanthids; some species have been dredged up apparently unassociated with other animals and are presumed to live free on the bottom. The associations appear to be very specific, for example, *Rhopalomenia aglaopheniae* always occurs associated with the hydroid *Aglaophenia*.

As in the Caudofoveata, remarkably little is known about the normal activities of this group presumably because of their scarcity and because they are normally only identified after collection and preservation; live specimens have rarely been studied. They are sluggish animals and may remain immobile for long periods but they can move along the surface of a secreted mucous thread by means of ciliary activity in the ventral groove; there appears to be no muscular activity involved in locomotion. They feed mainly on their cnidarian hosts and nematocysts are regularly found in the gut. The radula, when present, acts to comminute the food and is apparently used in a manner similar to that in gastropods. Solenogastres are hermaphrodite but it is generally held that they normally copulate and cross-fertilise; several authors have reported two individuals coiled round each other with the cloacal openings in contact and this is presumed to be copulation. The larval development is partially known for a few species only but it is presumed to include a planktonic phase. Nothing is known of their growth rates, longevity or sensory capabilities.

Collection and Preservation

Solenogastres are found associated with various sessile Cnidaria (Coelenterata) particularly hydrozoans and alcyonarians, which form at least a part of their food source. Some species are to be found on or in muds associated with such cnidarians while others may be found coiled around the hydroid stems, etc. If collected alive, they should be narcotised before fixation in non-acidic fixatives such as 70% alcohol or neutral 4% formaldehyde-seawater solution: this is to prevent destruction of the calcareous spicules necessary for identification. Examination of spicules will require access to a microscope and a good dissecting microscope will be needed even for a tentative identification.

Classification

Based upon Salvini-Plawen (1978)

Order PHOLIDOSKEPIA
- Family Dondersiidae (Simroth, 1893)
  - *Nematomenia banyulensis* (Pruvot, 1890)
- Family Lepidomeniidae (Pruvot, 1902)
  - *Lepidomenia hystrix* (Marion & Kowalevsky, 1886)

Order NEOMENIAMORPHA
- Family Neomeniidae (Ihering, 1867)
  - *Neomenia carinata* (Tullberg, 1875)
  - *Neomenia dalyelli* (Koren & Danielssen, 1877)

Order CAVIBELONIA
- Family Rhopalomeniidae (Nierstrasz & Stork, 1940)
  - *Rhopalomenia aglaopheniae* (Kowalevsky & Marion, 1887)
- Family Pararhopaliidae (Salvini-Plawen, 1972)
  - *Eleutheromenia sierra* (Pruvot, 1890)

Key

1. Cloacal gills present as a circlet of distinct radiating folds **2**

Cloacal gills absent **4**

2. Body with strongly developed dorsal keel broken into 15 or 16 compressed lobes of irregular size and shape; distich radula present; spicules hooked .... *Eleutheromenia sierra* (p. 36)

Short, thick, often semilunarly curved body; radula absent; spicules mainly open and channel shaped .... *Neomenia* (3)

3. Back keeled; channel-shaped dorsal spicules, many with arrowhead-shaped apices distinctly marked off from the shafts; length up to 30 mm .... *Neomenia carinata* (p. 30)

Back rounded; channel-shaped spicules with apices only a little broader than, and not markedly set off from, the shafts; length up to 20 mm .... *Neomenia dalyelli* (p. 32)

4. Hypodermis with papillae; body vermiform with convex sides, mostly circular in profile, not compressed; ventral fold or folds distinct; spicules needle-shaped, tangential and radial. Radula absent .... *Rhopalomenia aglaopheniae* (p. 34)

Hypodermis a single-celled layer without papillae; body vermiform but usually compressed with concave sides; ventral fold indistinct **5**

5. Radula absent; spicules mostly squamiform ....
.... *Nematomenia banyulensis* (p. 26)

Distich radula present; spicules pointed at least in part ....
.... *Lepidomenia hystrix* (p. 28)

Family DONDERSIIDAE Simroth, 1893

Mantle integument consisting of a moderate cuticle bearing two or more different types of solid calcareous scales. True stalked epidermal papillae are absent. Small animals up to about 20 mm long whose bodies may be uniformly cylindrical but are often tapering posteriorly. The radula has a median monoserial row of plates with a pair of arcuate, long, pointed denticles which are basally united; the radula is absent in a few forms. The mantle cavity lacks secondary respiratory folds and the dorsoterminal sensory organ may or may not be present. Hermaphrodite, but often lack copulatory armature and an unpaired secondary sexual orifice is present.

Genus NEMATOMENIA Simroth, 1893

Body usually laterally compressed and with the pedal groove indistinct. Body covered with a mixture of squamiform spicula and narrower, stalked ones. Radula and copulatory spines absent.

Type species: *Dondersia flavens* (Pruvot).

Nematomenia banyulensis (Pruvot, 1890)
(Fig. 6)

Dondersia banyulensis (Pruvot, 1891)
Myzonemia banyulensis (Simroth, 1893)

Greyish-red vermiform animals up to 30 mm long (Fig. 6A) but less than 1 mm in breadth; in section the sides of the body are slightly concave so that the body is taller than it is wide (Fig. 6B). The dorsal surface bears a distinct keel formed by erect spicules while the ventral furrow is indistinct and tapers towards the rear of the body until it is visible only as a stripe leading to the sub-terminal cloaca. Cloacal gills or copulatory organs are absent. A dorsal sensory organ is usually distinct as a wart-like projection on the dorsal surface just anterior to the cloaca. The pharynx is protractable but there is no radula present.

The body surface is covered by scale-shaped spicules, which are up to 0.25 mm long and are found on the lateral surfaces; these spicules are truncated at their bases (Fig. 6C, a). Narrower spicules (Fig. 6C, d) are mixed with the scale-like ones on the side of the body and in the region of the ventral furrow the scales are more lanceolate in shape (Fig. 6C, b & c).

This species has been subdivided into two subspecies, *N. banyulensis banyulensis* (Pruvot, 1890) which is a southern form found in the Mediterranean and along the Atlantic coast as far north as Plymouth and *N. banyulensis norvegica* Odhner, 1919 which is found in northern Europe. The latter form is distinguished by its smaller size (a maximum of about 15 mm long) and

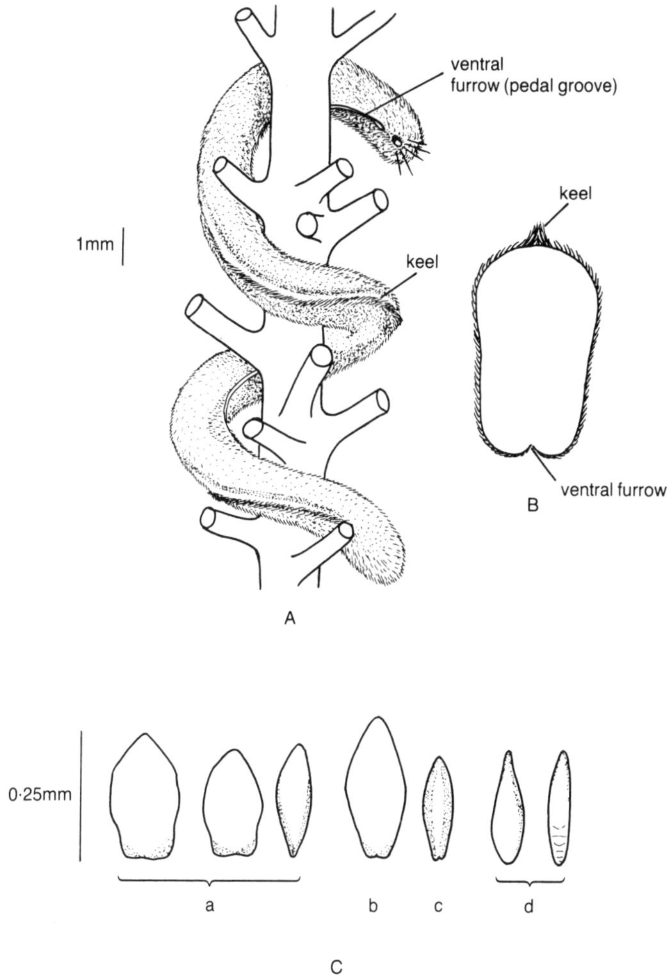

Fig. 6. *Nematomenia banyulensis*. A. External appearance of entire specimen in typical living position. B. Typical cross-section of body showing its slightly concave sides. C. Range of spicules found. (A and C after Salvini-Plawen 1971).

its more abruptly truncated spicule bases. They are found from depths of 45 to 300 m where they are epizoic on hydrozoans, principally *Lafoea dumosa, Lytocarpia myriophyllum* and *Grammaria abietina*, around whose branches they are coiled and upon which they are presumed to feed. In Britain, they have been recorded from the western coasts of England, Wales and Scotland and from the North Sea.

Family LEPIDOMENIIDAE Pruvot, 1902

Body usually slender and tapering posteriorly. The integument usually consists of a moderate cuticle lacking true stalked papillae and bearing a layer of solid calcareous scales externally: the scales which are also tapering in shape, increase in length towards the end of the body. The radula is distich, consisting of serially arranged, forceps-like denticulations. The mantle cavity usually lacks secondary respiratory folds. A dorso-terminal sense organ is usually present. They are hermaphrodite, with or without copulatory armature and an unpaired sexual orifice is present.

Genus LEPIDOMENIA Kowalevsky & Marion, 1887

Body tapering distinctly at the posterior end and rounded in section. Pedal groove indistinct. Body covered with slightly recurved, simple imbricate scales. Distich radula present. Secondary respiratory folds absent.

Type species: *Lepidomenia hystrix* (Marion and Kowalevsky).

Lepidomenia hystrix (Marion and Kowalevsky, 1886)
(Fig. 7)

Very small animals, only 2 mm long in a relaxed condition and which shorten by about 1/3 when contracted. The body tapers markedly towards the posterior end, the anterior region being about twice the diameter of the posterior region (Fig. 7). The body is conspicuously rounded in section throughout its length and the pedal groove runs from the mouth to the anus but is indistinct. The body is uniformly covered with strong, slightly recurved, tapering, simple imbricate scales, particularly evident on the lateral edges of the animal, and which are generally directed upwards and backwards. A distich radula is present bearing serially arranged, forceps-like denticulations. The mantle cavity lacks secondary respiratory structures.

This is a western Mediterranean form found in sand from shallow water to about 30 m and in the British Isles it has been recorded only from Belfast Lough, N. Ireland. In France and in the Mediterranean it has been found associated with the solitary coral *Balanophyllia* sp.

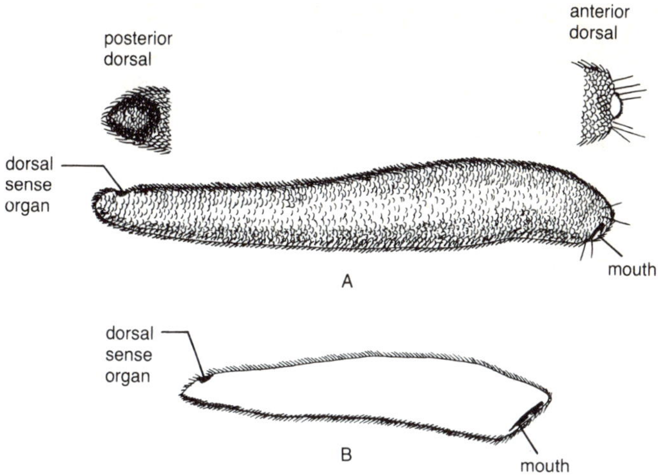

Fig. 7. *Lepidomenia hystrix*. A. External appearance of relaxed animal showing anterior dorsal view and posterior dorsal view. B. Profile of contracted animal. (After Kowalevsky and Marion 1887).

Family NEOMENIIDAE Ihering, 1867

Body short, truncated anteriorly and posteriorly and with or without a dorsal keel: it may be quite large, up to 100 mm long. Mantle integument with a strong cuticle and spicules comprising mainly long, slender and grooved ones together with solid, needle-like spines. Epithelial papillae present. Mouth opens into the atrium but there is no radula. A dorso-terminal sense organ is present and the mantle cavity has secondary respiratory folds. Hermaphrodite with copulatory armature consisting of a pair of stylets and an associated gland.

Genus NEOMENIA Tullberg, 1875

Semilunarly curved body with or without a keel. Radula absent. Gills a distinct circlet of radiating folds. Numerous ventral folds in the ventral furrow. Spicula mainly open and channel-shaped; hypodermis with broad cylindrical papillae.

Type species: *Neomenia carinata* Tullberg.

Neomenia carinata Tullberg, 1875
(Fig. 8A–C)

Solenopus nitidulus (Koren & Danielssen, 1877)

Greyish-white to pale pink animals with a short (up to 30 mm long), thick body truncated both anteriorally and posteriorally; dead animals frequently assume a crescent-shaped curve (Fig. 8A, B). The dorsal surface bears a characteristic low keel whose height is less than one fifth of the total body height. The ventral surface has a groove or furrow with numerous (7–9) folds. The terminal mouth is muscular with an extrusible pharynx but without a radula. The cloaca contains a circlet of some 40 distinct radiating gill folds. The body has a fairly uniformly distributed covering of small (up to 100 μm long), slender calcareous spicules which are characteristically open and channel-shaped (Fig. 8C); those from the region of the dorsal keel are largely shaped like arrowheads, the heads being sharply demarcated from the shafts.

They are found on soft or mixed bottoms in water from 18–565 m depth. They have been recorded from the North Sea and off the western coasts of Scotland, Ireland, Wales and England. They are sluggish animals normally found buried vertically in sediment with only the cloaca exposed, burrowing taking place using the evertible pharynx.

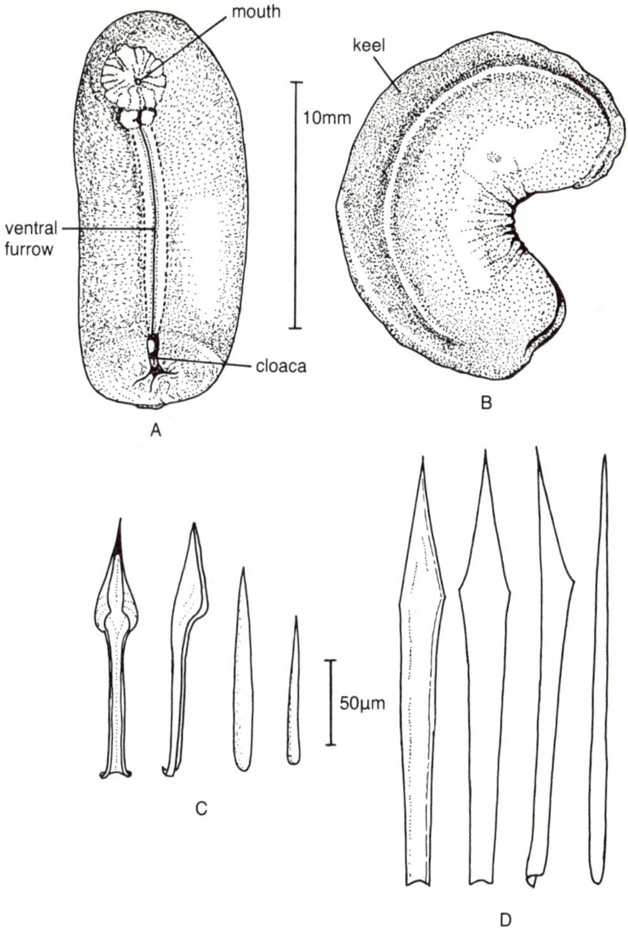

Fig. 8. A. Ventral view of *Neomenia carinata*. B. Lateral view of *N. carinata* showing crest. C. Dorsal spicules of *N. carinata* showing caniculate structure. D. Dorsal spicules of *Neomenia dalyelli*. (All after Muus 1959).

Neomenia dalyelli Koren and Danielssen, 1877
(Fig. 8D)

Solenopus dalyelli (Koren & Danielssen, 1877)

Similar in general appearance and structure to *N. carinata* except that the animals are smaller (up to 20 mm long) and lack any keel, the dorsal surface being rounded in outline. The caniculate spicules of the dorsal mid-line have apices only a little broader than, and not sharply demarcated from, the shafts (Fig. 8D). They have been recorded from the North Sea and the Shetland Islands where their habits and habitats appear similar to those of *N. carinata*.

Family RHOPALOMENIIDAE Nierstrasz & Stork, 1940

Body with a relatively smooth surface, the integument consisting of a thick cuticle pierced by stalked epidermal papillae and covered by several layers of hollow, needle-like calcareous spicules. The radula sometimes absent or, if present, it is distich, consisting of serially arranged, forceps-like denticulations. Secondary respiratory organs are absent in the mantle cavity. Hermaphrodite without copulatory armature but an unpaired secondary sexual orifice.

Genus RHOPALOMENIA Simroth, 1893

Body elongated but without a dorsal keel. Cuticle has thick epidermal papillae and well developed needle-like spicules. Without a radula but the salivary glands are massive. The midgut has serial lateral pouches. Copulatory armature absent.

Type species: *Rhopalomenia aglaopheniae* (Kowalevsky & Marion).

Rhopalomenia aglaopheniae (Kowalevsky & Marion, 1887) (Fig. 9)

Proneomenia aglaopheniae (Kowalevsky & Marion, 1887)

Elongate, vermiform animals up to 35 mm long, 3 mm wide and yellowish-grey in colour (Fig. 9A). The body is mainly circular in section but the sides may be slightly convex. The body lacks a dorsal keel; the distinct ventral groove is united with the fold; and the cloaca lacks secondary respiratory structures. The body surface appears fairly smooth, the thick cuticle being pierced by stalked, club-shaped papillae arising from the hypodermis and bearing several layers of hollow, needle-like calcareous spicules (Fig. 9B): hooked spicules and cloacal spines are absent. The radula is degenerate but salivary glands are well-developed.

They have been found off both the east and west coasts of Great Britain from depths of 50 m or more and they are usually associated with the hydroid *Lytocarpia (Theocarpus) myriophyllum* around which they are found coiled and twisted. This is a southern form found widely in the Mediterranean.

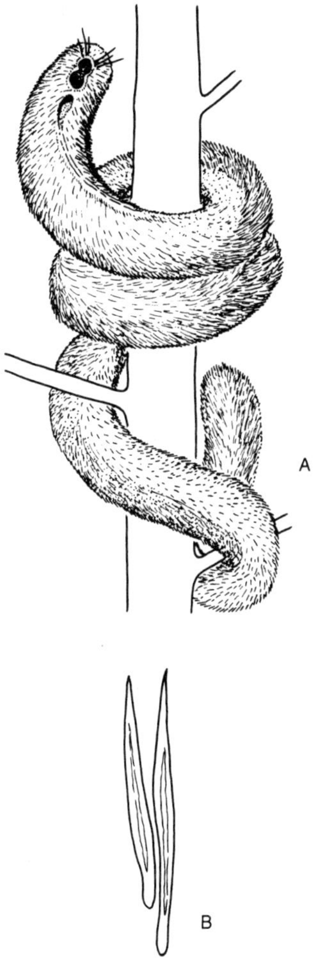

Fig. 9. *Rhopalomenia aglaopheniae*. A. General external appearance of adult in a typical living position. B. Typical body spicules. (After Salvini-Plawen 1971).

Family PARARHOPALIIDAE Salvini-Plawen, 1972

Body with a distinctly rough surface, the integument mainly bearing hollow, needle-like calcareous spicules in one or several layers; the thick cuticle is penetrated by stalked epidermal papillae. The spicules may sometimes be hooked. The radula is distich with serially arranged, forceps-like denticulations or is lacking. The mantle cavity may have secondary respiratory structures. Hermaphrodite, sometimes with seminal vesicles. A highly variable family.

Genus ELEUTHEROMENIA Salvini-Plawen, 1967

Characterised by the presence of hooked spicules, a distich radula, no copulatory spines but secondary gill folds are present. The ventral foregut glands are saclike.

Type species: *Paramenia sierra* Pruvot.

Eleutheromenia sierra (Pruvot, 1890)
(Fig. 10)

Paramenia sierra (Pruvot, 1890)

Small, yellowish-white animals up to 15 mm long and 2 mm broad (Fig. 10A). The dorsal surface has a strongly developed keel, broken up into 15 or 16 compressed lobes of irregular size and shape. A well-developed ventral groove runs from the distinct pedal pit to the terminal posterior cloacal opening which is circular but notched where the pedal groove joins it; the cloaca contains a circle of 28 retractable, ciliated branchial papillae. The body is covered by well-developed spicules which give the animal a distinctly 'hairy' appearance: those on the ventral and lateral surfaces are simple and needle-shaped (Fig. 10C) while the dorsal surface bears hooked spicules (Fig. 10B), each bearing a small, sharp point in the recurved area. The spicules are embedded in the thick cuticular covering. A distich radula is present and salivary glands are well-developed. Their ecology is unknown; the only British record is from a mixed bottom at 52 m depth in the Irish Sea while in other areas of Europe they have been found associated with hydroids such as *Aglaophenia myriophyllum*.

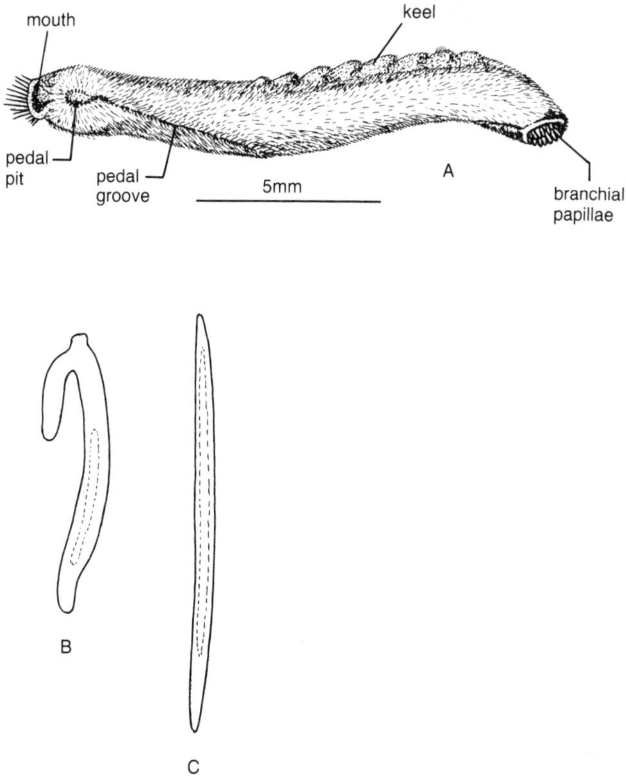

Fig. 10. *Eleutheromenia sierra*. A. General appearance. B. Hooked spicule from dorsal surface. C. Needle-shaped spicule typical of lateral and ventral surfaces. (After Salvini-Plawen 1971).

POLYPLACOPHORA

Introduction

The Polyplacophora are a group of entirely marine molluscs commonly known as chitons or 'coat-of-mail' shells. They are an ancient group, isolated valves having been found in the Upper Cambrian, but the fossil record is generally poor. Although conspicuous members of the littoral and sub-littoral fauna in many regions of the world, the British members tend to be drab and inconspicuous; only 11 (or possibly 12) species occur in the British Isles. The biology of the British species is poorly known largely because they have tended to be overlooked, being for the most part small in size, somewhat cryptic and predominantly sub-littoral in habit.

General Structure

Chitons are flattened dorso-ventrally and are bilaterally symmetrical (Fig. 11A, B). The **mantle** bears eight transverse **shell valves** on its dorsal surface (Fig. 11B), these valves articulating in such a way as to allow the animal to curl up into a ball when disturbed. Surrounding these valves is the **girdle** (Fig. 11A, B, C), a muscular flap which typically bears calcareous scales and/or spines; the girdle is a reflexed fold of the mantle tissue. Ventrally (Fig. 11C), the dominant feature is the large, flat, and muscular creeping **foot** which is surrounded by the **mantle** (or pallial) **groove** which contains numerous pairs of gills or **ctenidia** and into which opens the **anus**, the **renal pores** and the **genital apertures**. The anterior, centrally placed **head** is reduced and not clearly demarcated from the rest of the body.

Valve morphology is one of the principal taxonomic characters and the terminology used is summarised in Figs. 11 D–G. The **head valve** is typically undifferentiated (Fig. 11D). The dorsal surface of the **intermediate valves** can be divided into three main areas; a central **jugal area** flanked by two **median triangles** and posteriorly two **lateral triangles** (Fig. 11F). The **tail valve** is divided into two areas; the **antemucronal** and the **postmucronal areas** defined by the position of the **mucro** (Fig. 11E). The shell valves are complex structures composed typically of four layers: the thin, outer **periostracum** which contains most of the pigment; the **tegmentum** which is composed of **conchiolin** impregnated with calcium carbonate and which is perforated by numerous horizontal and vertical canals bearing the unique organs known as **aesthetes**; the **articulamentum** which is the thickest layer and is composed of crystalline calcium carbonate and the **hypostracum**, a thin innermost layer found only in some regions of the valves. The edges of the valves, where they are embedded in the girdle, are termed **insertion plates** (Fig. 11F), while the articulating projections found on the anterior edge of all valves except the head valve are termed the **sutural laminae** (Fig. 11F); both these structures are projections of the articulamentum. Notches occur in the insertion plates of some species (Fig. 11G); rows of holes (**notch rays**) perforating the ventral surface of the valve (Fig. 11G) run diagonally inwards from the notches towards the posterior edge of valves I–VII or towards the mucro of valve VIII. The jugal area of valves II–VIII may also be perforated by similar holes (Fig. 11G); all such holes are the points of entry of some of the **aesthete canals**. The sculpturing and morphology of the valves is closely related to the abundance and distribution of the aesthete canals which perforate the anterior and lateral edges, and sometimes the ventral surfaces, of some or all valves. The canals accommodate outgrowths of the mantle epithelium and run through the valves forming horizontal and/or vertical sections. These outgrowths terminate in apical sense organs, the aesthetes (Fig. 12A) whose precise function/s remains uncertain but which is thought to be related to light reception in some species at least. Each aesthete complex is, in some species, associated with a raised papilla on the valve surface (e.g. *Acanthochitona* spp) while in others the aesthete complexes run together without any sculpturing to separate them (e.g. *Callochiton* sp). Plate 1 illustrates the principal types of sculpturing and their association with the external endings (apical and subsidiary caps) of aesthete complexes. The valves are embedded in, and are secreted by, the mantle which forms the tough flexible girdle which runs around the margin of the animal. The dorsal surface of the girdle bears scales (or tubercles), protuberant chitinous or calcareous spicules, and sometimes compact tufts of bristles (Plate 2).

The mantle cavity encircles the animal, bounded internally by the foot and externally by the girdle. The anterior portion is occupied by the head (Fig. 11C) which is simple and bears no eyes or tentacles but has a median mouth armed with a **radula** whose chitinous teeth (Fig. 12B, C) may be used in laboratory confirmation of identifications (Plate 3). Sensory organs, the **osphradia**, lie in the posterior region of the mantle cavity but their sensory function is uncertain. The anus (Fig. 11C) opens into the mantle cavity posteriorally on a small median papilla: two renal pores open into the mantle cavity posteriorally, one on each side, as do the two gonopores which lie slightly anterior to the renal openings (Fig. 11C). The dominant structures of the mantle cavity, however, are the gills or ctenidia which are variable in number and lie in a postero-lateral position (Fig. 11C); two conditions are recognised, the **holobranch** condition where the ctenidia extend forward to the head (Fig. 11C1) and the **merobranch** condition where the ctenidia only extend part of the way to the head (Fig. 11C2). The basic structure and ciliation of the ctenidia are comparable to those of primitive gastropods but they are generally short and broad, lacking any internal chitinous supports.

The broad, flat foot is adapted for creeping over irregular rocky surfaces and for gripping the rock surface suctorially as a protection against disturbance.

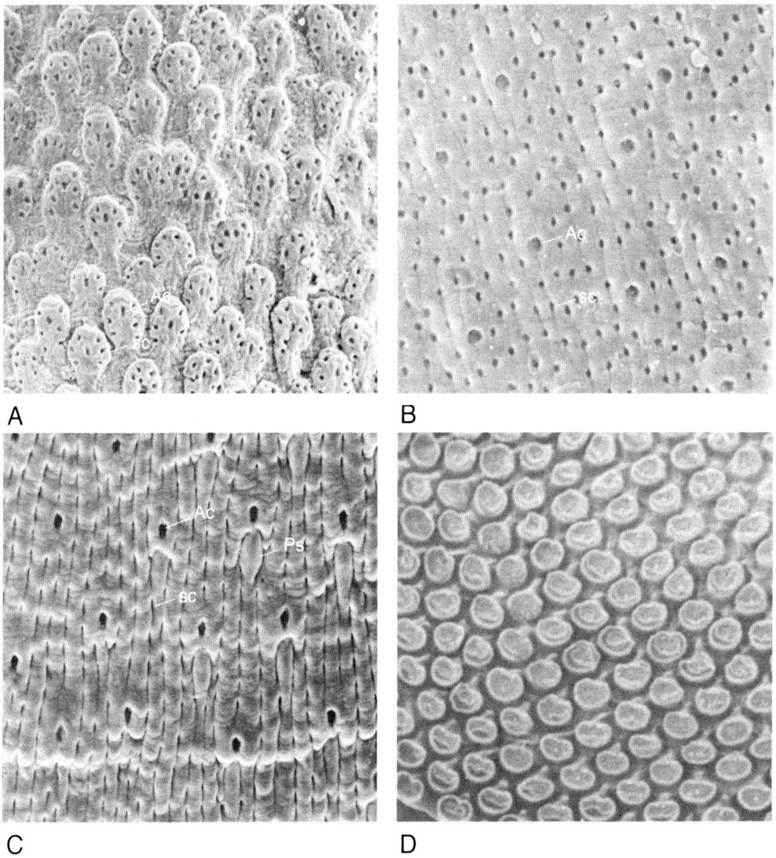

Plate 1. Principal types of sculpturing of the dorsal surface of the chiton shell valve. A. *Hanleya hanleyi:* Raised papillae, randomly arranged, each bearing an apical cap and associated subsidiary caps (×250). B. *Tonicella marmorea:* No obvious raised papillae; the apical caps are regularly arranged with subsidiary caps evenly distributed over the valve surface (×600). C. *Callochiton achatinus:* Furrowed valve surface with apical caps situated on small, very slightly raised, papillae, and the swellings associated with the pigment spots lying anterior to these. The subsidiary caps are arranged in precise lines (×750). D. *Acanthochitona fascicularis:* Valve surface covered by densely packed, round, raised papillae (×160). (Ac, apical cap; Ps, pigment spot swelling; sc, subsidiary cap).

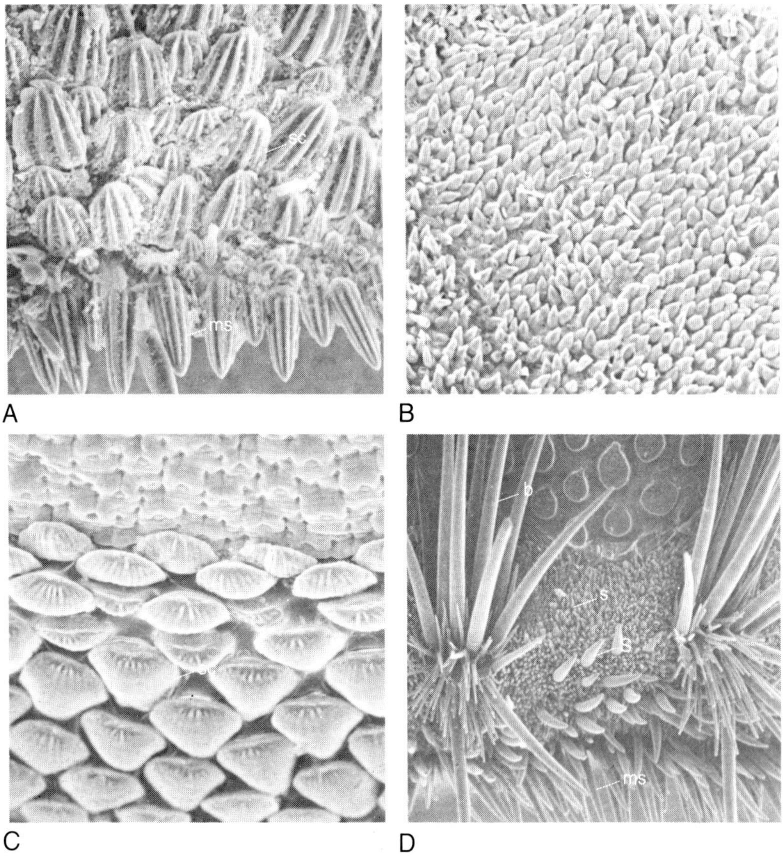

Plate 2. Principal types of girdle spiculation in chitons. A. *Leptochiton asellus:* Large, loosely packed scales bearing a number of grooves on the dorsal surface. The girdle is fringed by short, bluntly pointed, grooved marginal spines (×500). B. *Tonicella marmorea:* Girdle covered by densely packed, minute pointed granules (×300). C. *Ischnochiton albus:* Girdle covered by regularly arranged, very large scales covering the entire girdle (×500). D. *Acanthochitona crinitus:* Girdle bearing tufts of long, smooth bristles. Remainder of girdle covered by densely packed, small pointed spines with larger, slightly curved, grooved spines interspersed amongst them but concentrated towards the outer half of the girdle. Girdle fringed by a dense row of long, marginal spines (×100). (b, bristle; g, granule; ms, marginal spine; sc, scale; s, small spine; S, large spine).

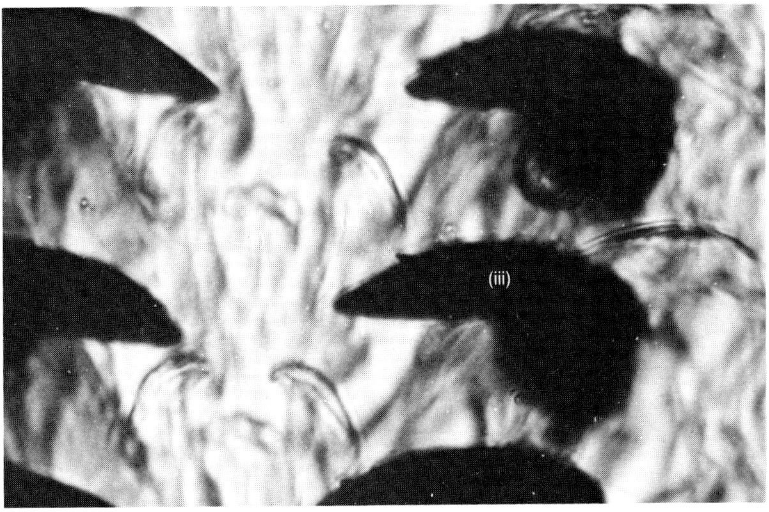

A

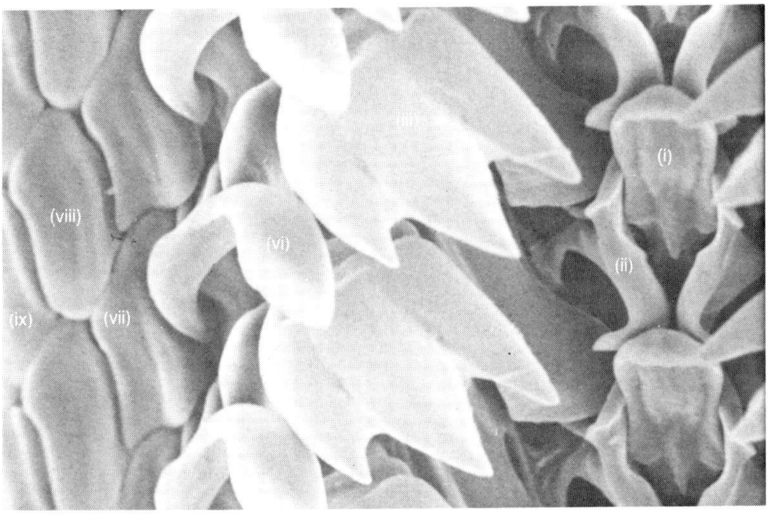

B

Plate 3. A. View of the radula of *Lepidochitona cinereus* as seen under the light microscope (×500). B. View of the radula of *Acanthochitona crinitus* as seen under the Scanning Electron Microscope (×500). ((i), median tooth; (ii), intermediate tooth; (iii), uncinate blade; (vi)–(ix), marginal teeth).

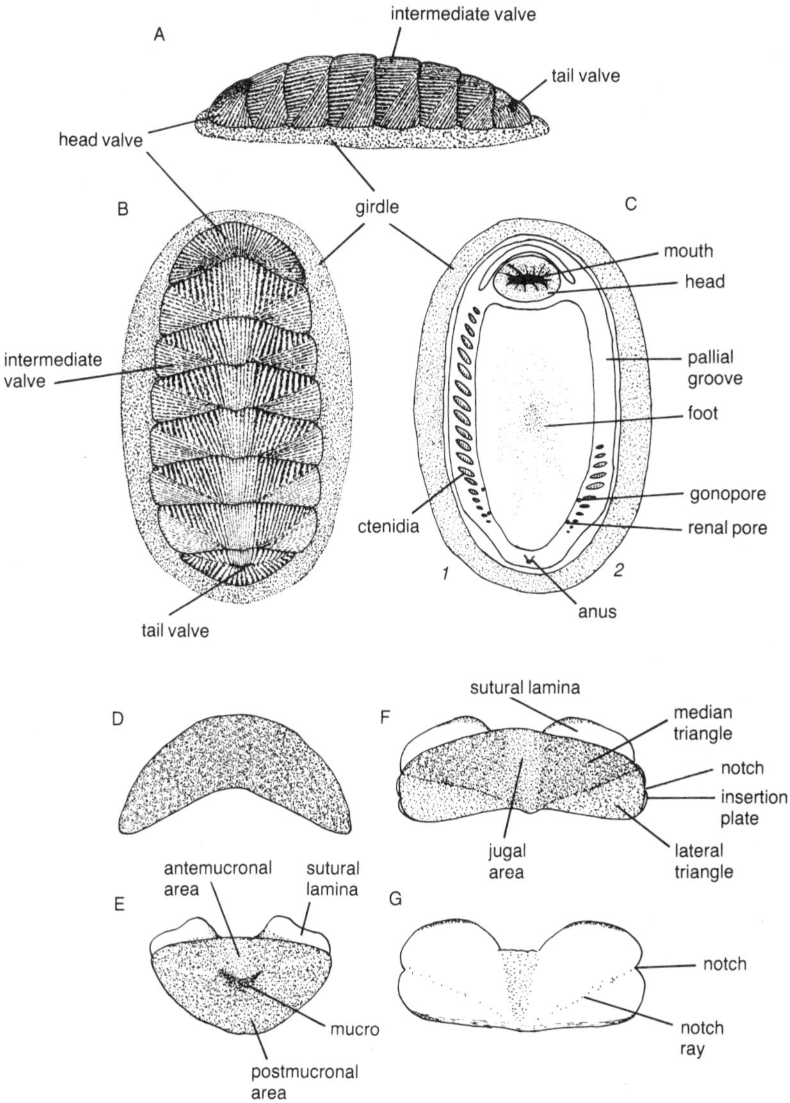

Fig. 11. General features of the Polyplacophora. A. Lateral view. B. Dorsal view. C. Ventral view: 1, Holobranch condition; 2, Merobranch condition. D. Dorsal view of head valve. E. Dorsal view of tail valve. F. Dorsal view of intermediate valve. G. Ventral view of intermediate valve.

Biology and Ecology

Chitons are world-wide in distribution but each species has its own characteristic geographical distribution and habitat range. They are important members of littoral and shallow sea communities in some regions and although the British species are small and inconspicuous, they are ecologically significant members of some sub-littoral habitats where they may be found in large numbers.

They are generally slow-moving animals and use the broad creeping foot for attachment and for movement over the often irregular hard surfaces on which they are found. All chitons, except for the non-British genus *Placiphorella*, are grazers/browsers of rock surfaces, feeding by scraping organic matter from the surface using the toothed radula which may be quite long but most of which is contained within the radula sac in the buccal cavity (Fig. 12C). Below the radula sac is the subradular sac which contains a chemoreceptor, the **subradular organ** (Fig. 12C). During feeding, the subradular organ is extended out of the mouth periodically to test the substrate; if food is detected, it is withdrawn and the radula is protruded and is used to scrape material from the surface. Mucus from the salivary gland lubricates the radula and catches small particles as they are removed by the radula teeth; this mucus then forms strings containing the food particles which are then passed on into the digestive system. The feeding action is readily observed through the walls of a glass aquarium.

Although some species of chitons grow up to about 300 mm in length, the British representatives are generally small, rarely exceeding 40 mm in length. Shell growth takes place on the anterior and lateral edges of the head and intermediate valves and on all edges of the tail valve; such new growth is often clearly visible and in some species, such as *Lepidochitona cinereus*, the new growth may be of a distinctly different colour. Most species exhibit characteristic colours and/or patterns of the shell valves and the girdle: however, some species such as *L. cinereus* are extremely variable in both colour and pattern.

Chitons are **dioecious** but the sex of an individual can only be determined by dissection. Reproduction in most species is markedly seasonal and fertilisation is external, both eggs and sperm being released into the seawater. The eggs are deposited singly, in masses or in strings and there is a veliger larval stage absent, the larval **trochophore** metamorphosing directly into a juvenile; the trochophore is usually planktonic.

The principal predators of chitons appear to be fish and crabs which may be able to prise the chiton from the rock surface; when dislodged, they frequently curl into a ball which effectively protects their soft tissues. Chitons are animals of hard surfaces (rocks, stones, shell debris, etc) and the littoral British species are usually found under stones, often those partially embedded in sand or gravel, or in dark crevices. Chitons are known to occur from the mid-littoral down to depths of 400 m but they occur mainly in the lower littoral and shallow sub-littoral.

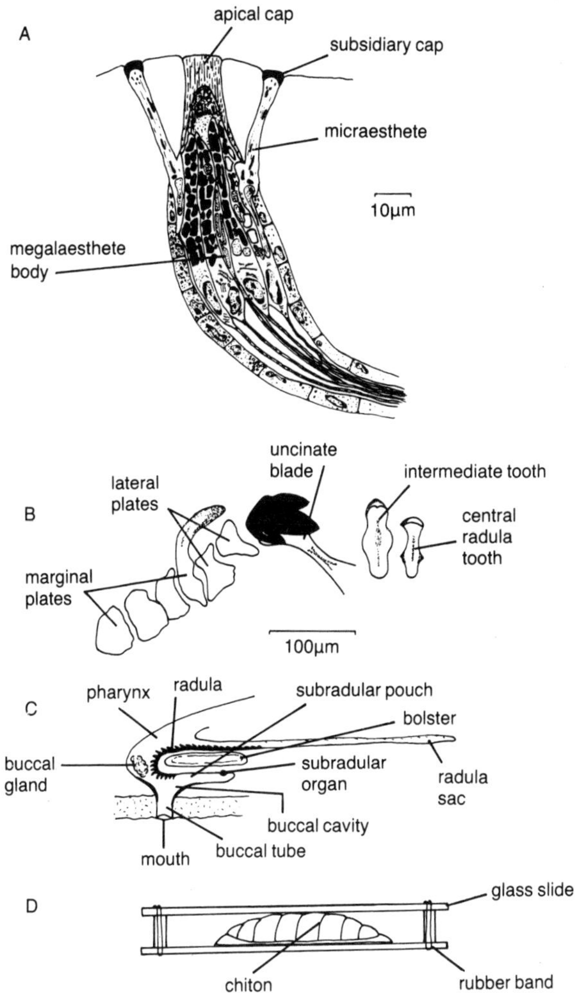

Fig. 12. General features of the Polyplacophora. A. Generalized structure of an aesthete. B. Half of one transverse row of radular teeth. C. Longitudinal section through the buccal cavity. D. Method of maintaining a flat specimen during the fixing precess.

Collection and Preservation

Littoral species can be collected most readily from beneath stones and boulders below M.L.W.N. tide level on semi-exposed and sheltered rocky shores. Sub-littoral species are collected from dredged or trawled material, either stones or shell debris or by SCUBA divers. They can be removed easily from smooth surfaces by pushing them carefully sideways but from irregular surfaces they must be prised off very carefully by inserting a blade beneath the girdle and lifting or twisting. Specimens will curl up on being removed and should be placed in a rigid container, preferably without right-angled corners to aid in subsequent removal, containing only a minimum of seawater: they can be encouraged to re-attach by a gentle but rapid tapping of the rim of the container.

Specimens for routine collections can be fixed in 70% alcohol, 4% neutral formaldehyde/seawater solution or they can be air-dried; it should be noted that alcohol preservation sometimes removes some of the pigmentation from the valves and girdle. Animals should be killed in a flattened position which is best achieved either by narcotising the animals while attached normally to a flat surface or by holding the animal flat while the killing and fixation processes are being carried out. Narcotisation may be achieved by any of the normal methods applicable to the Mollusca and including the use of magnesium sulphate or magnesium chloride solutions or menthol crystals placed on the surface of the water; this is a slow process, however. The best method is to contain the animal between two glass slides as shown in Fig. 12D and then to immerse the system in the fixative: the glass slides may be removed after about an hour to allow optimal fixation of the animal.

The examination of the radula and/or the ventral and lateral surfaces of the valves for laboratory confirmation of identifications clearly requires a degree of damage to be inflicted on the specimen. The methods used for killing and fixing should be chosen with this in mind. The radula can often be withdrawn from the buccal cavity of a freshly killed animal using a pair of fine forceps or by dissection if it has been decided to air-dry the specimen (when total removal of the foot and visceral mass is required anyway). Examination of valves is best carried out after either quickly killing the specimen by sudden immersion in boiling water for a few seconds (which makes valve removal very easy) or by dissolving the soft tissues of a dead animal away by immersing in a saturated solution of potassium hydroxide for several hours: such treatments result in the effective destruction of the specimen from a collectors viewpoint but may be necessary for unequivocal identification of some species. It is very difficult to remove individual valves by dissection from alcohol or formaldehyde-fixed specimens or from air dried specimens without damaging the valves in the process.

Examination in the field will usually require a hand lens while laboratory confirmations involving radula characteristics will require at least a good dissecting microscope.

Classification

Linnaeus (1758) first described and classified the 'coat-of-mail' animals by collecting them into the single genus *Chiton*. The term chiton has become accepted as the vernacular name for all members of this group although the genus *Chiton* is now confined to several species in the Caribbean.

The taxonomy of chitons was widely debated for most of the 19th century until Pilsbry (1892) presented a scheme of classification which forms the basic framework of the two modern classifications. Thiele (1929) divided chitons into two orders and six families based on shell, girdle and radula characters while Bergenhayn (1955) proposed an expanded classification to include fossil groups and comprised two wholly extinct orders and four others with a total of 14 families. It is the systematic classification of Bergenhayn that has been adopted by the authors for this *Synopsis* as it is not only the most recent but also the most complete work available.

The 12 species of British Polyplacophora are distributed among three orders.

Class POLYPLACOPHORA de Blainville, 1816
- Subclass NEOLORICATA Bergenhayn, 1955
  - Order LEPIDOPLEURINA Thiele, 1910
    - Family Lepidopleuridae Pilsbry, 1892
      - Genus *Leptochiton* Gray, 1847
        - *L. asellus* (Gmelin, 1791)
        - *L. cancellatus* (Sowerby, 1840)
        - *L. scabridus* (Jeffreys, 1880)
    - Family Hanleyidae Bergenhayn, 1955
      - Genus *Hanleya* Gray, 1857
        - *H. hanleyi* (Bean in Thorpe, 1844)
  - Order ISCHNOCHITONINA Bergenhayn, 1930
    - Family Ischnochitonidae Dall, 1889
      - Sub-family Lepidochitoninae Iredale, 1914
        - Genus *Lepidochitona* Gray, 1821
          - *L. cinereus* (Linnaeus, 1767)
        - Genus *Tonicella* Carpenter, 1873
          - *T. rubra* (Linnaeus, 1767)
          - *T. marmorea* (Fabricius, 1780)
      - Sub-family Ischnochitoninae Dall, 1889
        - Genus *Ischnochiton* Gray, 1847
          - Sub-genus *Stenosemus* Middendorff, 1847
            - *I. (Stenosemus) albus* (Linnaeus, 1767)
      - Sub-family Callochitoninae Plate, 1901
        - Genus *Callochiton* Gray, 1847
          - *C. achatinus* Brown, 1823

Order ACANTHOCHITONINA Bergenhayn, 1930
Family Acanthochitonidae Pilsbry, 1893
Sub-family Acanthochitoninae Pilsbry, 1893
Genus *Acanthochitona* Gray, 1821
A. fascicularis (Linnaeus, 1767)
A. crinitus (Pennant, 1777)
A. discrepans (Brown, 1827)

Field Key

The key to the British species of chiton (Matthews, 1953) uses the presence or absence of notches in the insertion plates of the various valves. In order to determine this characteristic it is necessary to dissect areas of the girdle which results in the death of the specimen. This is obviously an undesirable procedure and it is proposed to present here a non-destructive field key together with more exhaustive confirmatory laboratory notes for each species.

Girdle spiculation, valve sculpturing and ctenidial arrangement are the main characters used in the field key.

1. Girdle with 18 *tufts* of spines **2**
Girdle without *tufts* of spines **3**

2. Valve granulations *minute, rounded* shape, evenly and very densely distributed .... *Acanthochitona fascicularis* (p. 84)
Valve granulations *large, tear-drop* shaped, unevenly and sparsely distributed .... *Acanthochitona crinitus* (p. 88)
Valve granulations moderately sized, oval shaped regularly distributed .... *Acanthochitona discrepans* (p. 90)

3. Intermediate valves with *divaricate* and *linear* sculpturing, usually white/cream in colour but may be covered by an opaque black deposition **4**
Intermediate valves with only *divaricate* sculpturing **7**

4. Intermediate valves *rounded* and without *keel* **5**
Intermediate valves with definite *keel* **6**

5. *Lateral triangle* very narrow, *linear* sculpturing of *median triangle* arranged in long chain-like rows .... *Leptochiton cancellatus* (p. 54)
Linear sculpturing very coarse giving the valve surface a rough appearance .... *Leptochiton scabridus* (p. 56)

6. Girdle with sparsely distributed *single* spines occurring particularly in the area between adjacent intermediate valves .... *Hanleya hanleyi* (p. 58)
Girdle without spines .... *Leptochiton asellus* (p. 50)

7. Girdle covered in *large acicular* uniformly arranged spines. Valves are mottled brown, yellow and red with fine sculpturing. Dark *pigmented* megalaesthete bodies visible on head valve, *lateral triangles* of intermediate valves and *posterior area* of tail valve .... *Callochiton achatinus* (p. 78)

Girdle and valve ornamentation absent or faint, valves mottled red, pink and cream with a smooth appearance **8**

Girdle and valve ornamentation coarse **9**

8. Girdle is 'leathery', bearing *minute, widely* spaced granules .... *Tonicella marmorea* (p. 70)

Girdle *densely* covered in *small circular* granules, coloured in pink and cream bands. Valves are almost smooth and more often pink and cream rather than red .... *Tonicella rubra* (p. 66)

9. Granulations on girdle and sculpturing on valves equally prominent. Girdle often having a two-tone banding with cigar-shaped peripheral spines. Valves of very variable colour (black, white, green, pink, orange, blue) and pattern .... *Lepidochitona cinereus* (p. 62)

Granulations on girdle very prominent, *large oval* granules; the sculpturing on the valves is less pronounced. Valves white or cream .... *Ischnochiton albus* (p. 74)

Order LEPIDOPLEURINA Thiele, 1910

This order contains the more primitive families of extant polyplacophorans. The head valve is undifferentiated, whilst the intermediate and tail valves are differentiated into various areas. Insertion plates are lacking or very much reduced and never bear notches. The sutural laminae are small. All members of this order are merobranchial.

Family LEPIDOPLEURIDAE Pilsbry, 1892

Body relatively small, elongate and mostly less than 50 mm in length. Valves are well-developed but fragile and sculptured with densely arranged dorsal papillae. Insertion plates are lacking and the articulamentum is represented by slight sutural laminae on valves II–VIII. The valves are naturally white or beige in colour but may have deposits of iron or manganese on their surface imparting a yellow, orange or black colouration. The girdle is narrow and covered by minute spicules or scales and surrounded by a fringe of short bluntly pointed spines. Members of this family are typically merobranch. They occur from the littoral zone down to depths of greater than 7000 m. The family contains four genera, only one of which has representatives in British waters; these were until recently incorrectly classified in the type genus *Lepidopleurus* Risso, 1826 (Type species: *Chiton cajetanus* Poli, 1791). This is clearly incorrect and the nomenclature used here is that used by Kaas (1981) who correctly attributed these species to the Genus *Leptochiton* Gray, 1847.

Genus LEPTOCHITON Gray, 1847

Typically ovate to elongate. The valves are thin and fragile, evenly granulated and with only indistinctly defined, hardly elevated lateral areas. The girdle is narrow and densely covered by calcareous scales and spicules.

Type species: *Chiton cinereus* Montagu, 1803; SD Gray, 1847. (=*Chiton asellus* Gmelin, 1791); non *Chiton cinereus* Linnaeus, 1767.

Leptochiton asellus (Gmelin, 1791)
(Fig. 13)

Chiton asellus Gmelin, 1791
Chiton cimicinus Spengler, 1797
Chiton cinereus; Montagu, 1803
Chiton minimus; Wood, 1815
Chiton fuscatus Leach, 1852
Chiton scoticus Leach, 1852
Lepidopleurus asellus; Leloup, 1934
Leptochiton asellus; Kaas, 1981

A merobranch chiton with a distinctly oval shape (breadth : length = 0.6–0.7) and with a recorded maximum length of 19 mm. The girdle is narrow with a shell width : total width ratio of about 0.8 (Fig. 13A).

The shell valves are thin and fragile and lack insertion plates. Valves II–VIII have small, approximately triangular sutural laminae (Fig. 13F, G). The intermediate valves are of medium height with almost straight sided slopes, definitely although only moderately keeled (Fig. 13B), and typically with a beak particularly on valves II and III, although this is often absent due to erosion of the posterior edge of the valves (Fig. 13F). The tail valve is rounded-triangular with a relatively prominent mucro situated slightly anteriorly (Fig. 13G).

The dorsal surface of the valves is granular when seen under a hand-lens; the intermediate valves have large, slightly elevated, conspicuous lateral areas as defined by the orientation of the granules (Fig. 13F). Examination under the dissecting microscope reveals the granular sculpturing in the jugal and median areas to be due to slightly disjointed, linear rows of roughly oval dorsal papillae running strictly antero-posteriorly (Fig. 13F), and in the lateral area to be due to similar papillae running in diverging rows (Fig. 13F). Each dorsal papilla bears an apical cap situated towards its anterior end with only a few associated subsidiary caps on the posterior area of the papilla. Growth interruption lines are often clearly visible on all valves. The ventral surface of all valves is perforated by a few rows of holes close to their growing edges (Fig. 13H). The natural shell colouration is white or cream but deposits of iron/manganese concretions often produce streaks or patches varying from rust to opaque black in colour.

The girdle is narrow and is covered by densely packed, elongate, bluntly pointed scales (140 μm long) with 8–10 elevated riblets (Fig. 13C) and is bordered by a fringe of slender spicules (110μm long) (Fig. 13D).

The median radular plate has a weak central spine and wings which develop into a small curved blade at its distal end; the intermediate plates are slightly longer than the median plate and end in a similar sized curved blade (Fig. 13I). The uncinate blade has one strong central denticle and a second smaller one exteriorly (Fig. 13J).

L. asellus is common from around or just below low water mark to a

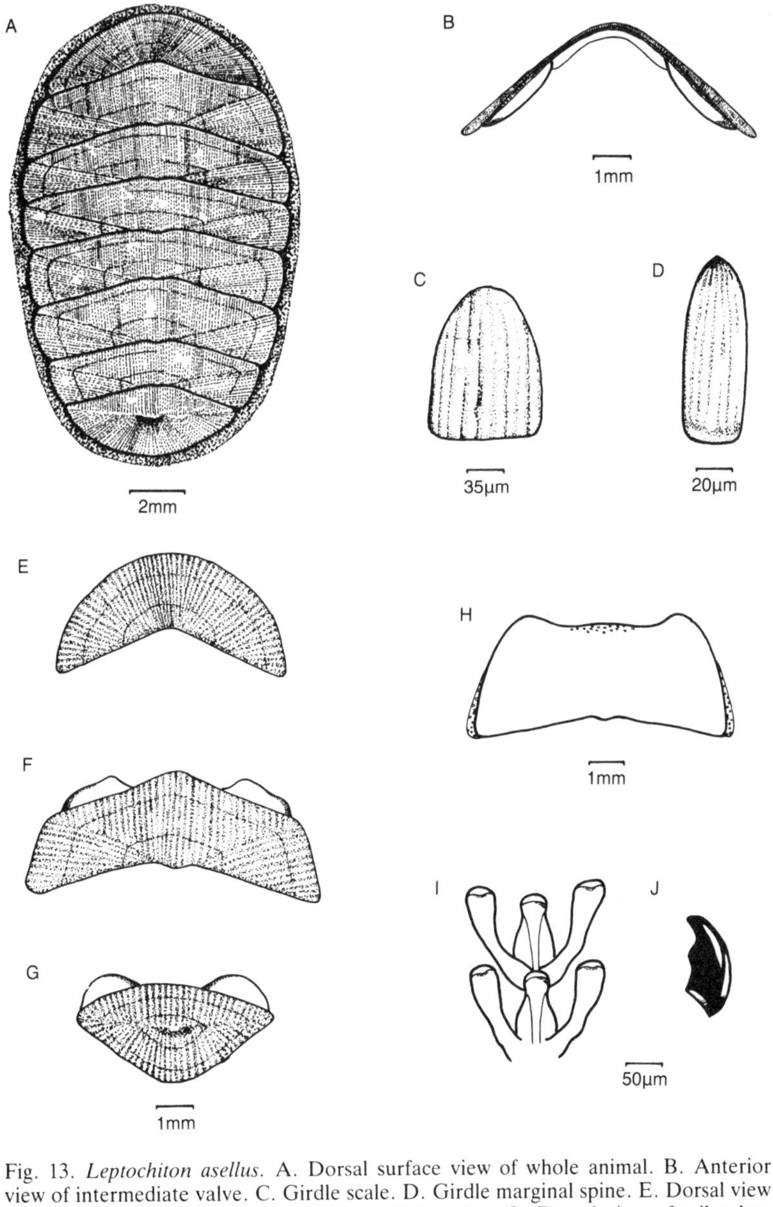

Fig. 13. *Leptochiton asellus*. A. Dorsal surface view of whole animal. B. Anterior view of intermediate valve. C. Girdle scale. D. Girdle marginal spine. E. Dorsal view of head valve. F. Dorsal view of intermediate valve. G. Dorsal view of tail valve. H. Ventral view of intermediate valve. I. Median and intermediate radula plates. J. Uncinate blade of radula.

depth of 250 m. It is found on silty-stony bottoms, often attached to shell debris. It is a very common northeast Atlantic species with a range extending from Spitsbergen and the Barents Sea in the north, Iceland and perhaps Greenland, to the British Isles and the Bay of Biscay as far south as Vigo, Spain. It is found off the entire coastline of the British Isles in large numbers.

Leptochiton cancellatus (Sowerby, 1840)
(Fig. 14)

Chiton cancellatus Sowerby II, 1840
Chiton tuberculatus Leach, 1852
Lepidopleurus cancellatus; Leloup, 1934
Leptochiton cancellatus; Ferreira, 1979

A merobranch chiton with an oblong shape (breadth : length ratio of 0.5–0.6) and with a recorded maximum length of 9 mm. The girdle is narrow with a shell width : total width ratio of 0.8–0.9 (Fig. 14A).

The shell valves are thin and fragile and lacking insertion plates. Valves II–VIII have very small, irregularly triangular sutural laminae (Fig. 14F, G). The intermediate valves are relatively tall, evenly rounded and arched, bearing neither a keel (Fig. 14B) nor a beak (Fig. 14F). The tail valve is relatively large, posteriorly rounded and bears a mucro which is relatively posteriorly located (Fig. 14G).

The dorsal surface of the valves is slightly granular in appearance when seen under a hand lens; the intermediate valves have very slightly elevated lateral areas defined by variations in the orientation of the granulations (Fig. 14F). Examination under the dissecting microscope reveals the granular sculpturing in the jugal and median areas to be due to regular, slightly converging rows of oval papillae running antero-posteriorly; the papillae are so closely packed they are almost touching, resulting in the smooth appearance of the surface. The lateral area is covered by noticeably more irregularly distributed papillae arranged in a diverging pattern. Each dorsal papilla bears an apical cap situated near its anterior end with only a few associated subsidiary caps on the posterior region. Growth interruption lines are often clearly visible on all valves. The natural shell colouration is white or cream but this can vary from rust to black in patches, streaks or all over due to deposits of iron/manganese concretions.

The girdle is covered by densely packed squarish scales (50–60 μm long) with 12–16 sharp, narrow riblets and scroll-like tips (Fig. 14C) and is bordered by a fringe of ribbed, pointed spicules (100 μm long) (Fig. 14D).

The median radular plate is dilated in the middle, tapering markedly at the posterior end and with a well-developed blade; the intermediate plates are narrowly wing-like without a blade (Fig. 14I). The uncinate blade has one strong central denticle and a second smaller one exteriorly (Fig. 14J).

L. cancellatus is not common and found around or just below low water mark, extending downwards to depths in excess of 900 m. It is found on silty-sandy bottoms often attached to small stones or shell debris. It has a restricted distribution, being found around the coast of the British Isles, the Atlantic coasts of France, Spain and Portugal and the Mediterranean.

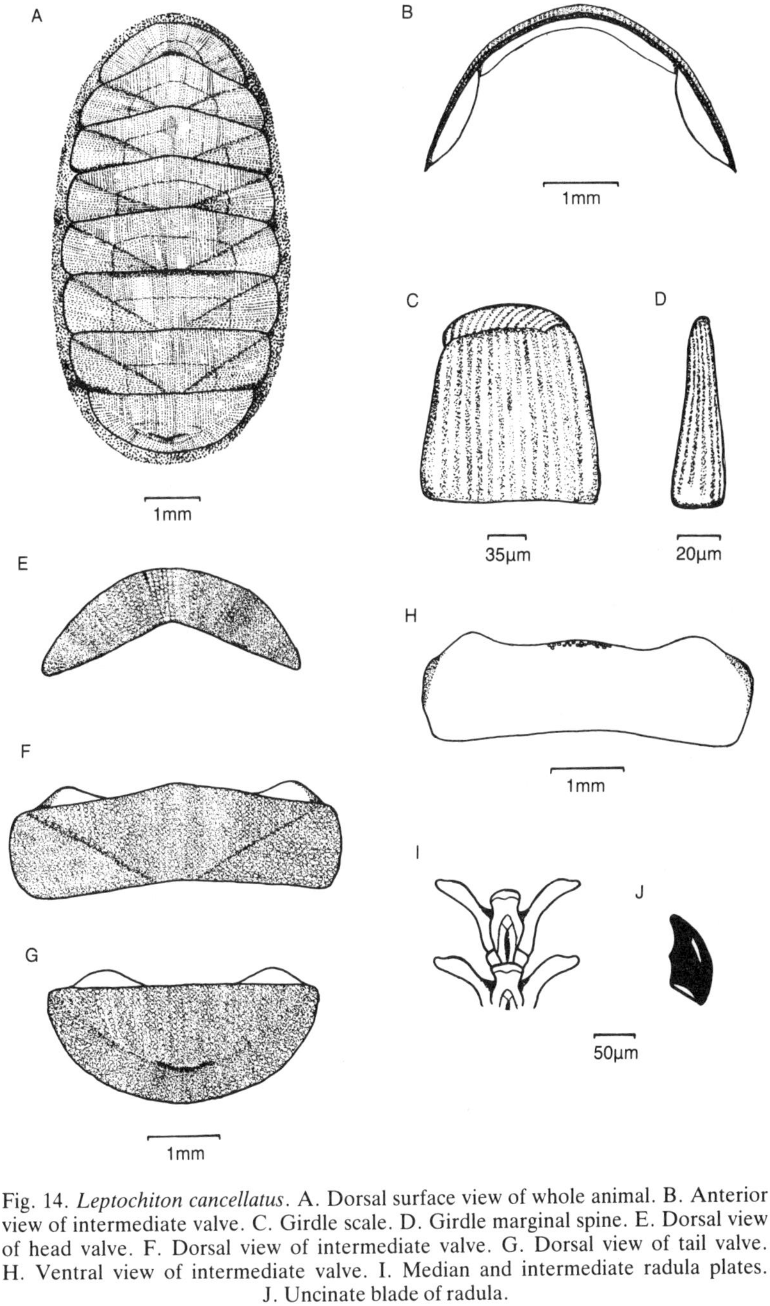

Fig. 14. *Leptochiton cancellatus*. A. Dorsal surface view of whole animal. B. Anterior view of intermediate valve. C. Girdle scale. D. Girdle marginal spine. E. Dorsal view of head valve. F. Dorsal view of intermediate valve. G. Dorsal view of tail valve. H. Ventral view of intermediate valve. I. Median and intermediate radula plates. J. Uncinate blade of radula.

Leptochiton scabridus (Jeffreys, 1880)
(Fig. 15)

Chiton scabridus Jeffreys, 1880
Lepidopleurus scabridus, Sykes, 1894

A merobranch chiton with an oblong shape (breadth : length ratio of about 0.5), and with a recorded maximum length of 8 mm. The girdle is narrow with a shell width : total width ratio of about 0.85 (Fig. 15A).

The shell valves are fragile, lacking insertion plates. Valves II–VIII have small, triangular sutural laminae (Fig. 15F, G). The intermediate valves are rounded and relatively flat with neither an obvious keel (Fig. 15B) nor a beak (Fig. 15F). The tail valve is semi-circular with a centrally located mucro (Fig. 15G).

The dorsal surface of the valves is granular when seen under a hand lens; the lateral areas of the intermediate valves are defined by the orientation of the sculpturing (Fig. 15F). Examination under the dissecting microscope reveals the sculpturing to be due to regular rows of raised, round dorsal papillae. Growth interruption lines are often visible. Shell colouration ranges from off-white to a dull orange and occasionally there are black deposits of manganese or iron. The foot of *L. scabridus* is a striking bright red colour in live specimens.

The girdle is covered by densely packed squarish scales (80–90 μm long) with 15–18 narrow riblets (Fig. 15C). It is bordered by a fringe of ribbed, pointed spicules (100 μm long) (Fig. 15D).

The median radula plate is broadest at its anterior end where it forms a well-developed blade (Fig. 15I); the intermediate plates are long and wing-like. The uncinate blade has three denticles, the central one being much the largest (Fig. 15J).

L. scabridus is rare, found at extreme low water and sublittorally. It is found attached to stones or pebbles often very heavily embedded in sand to a depth of 10–20 cm. It has a very restricted distribution being found only in SW England, the Channel Islands and the Britanny coast.

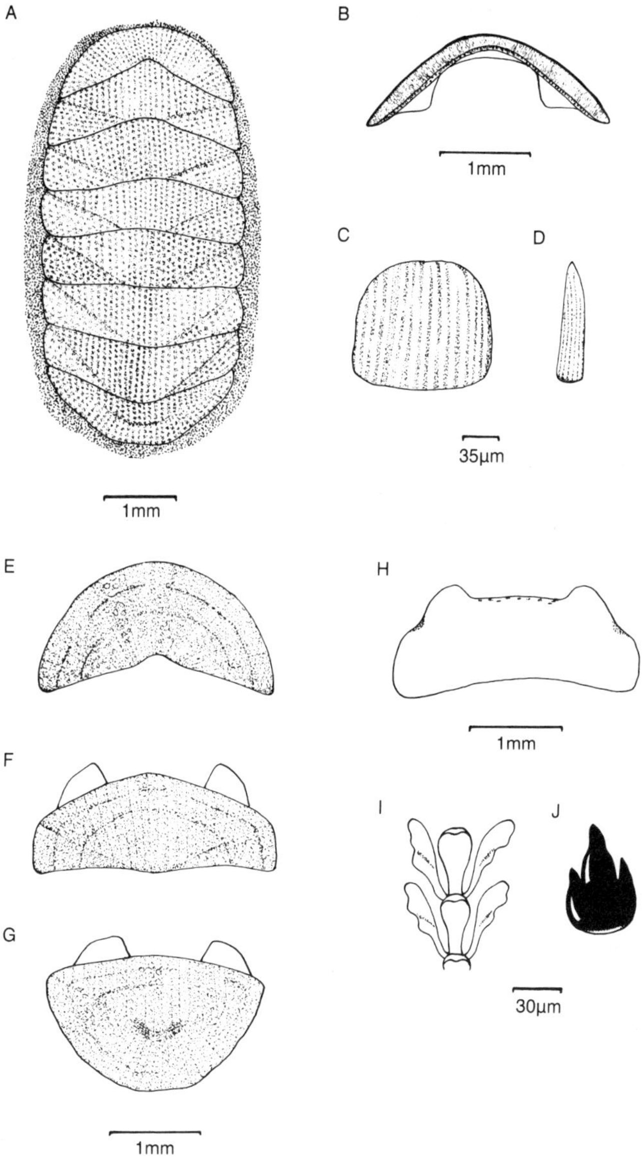

Fig. 15. *Leptochiton scabridus*. A. Dorsal surface view of whole animal. B. Anterior view of intermediate valve. C. Girdle scale. D. Girdle marginal spine. E. Dorsal view of head valve. F. Dorsal view of intermediate valve. G. Dorsal view of tail valve. H. Ventral view of intermediate valve. I. Median and intermediate radula plates. J. Uncinate blade of radula.

Family HANLEYIDAE Bergenhayn, 1955

Body ovate to elongate, largest specimens up to 60 mm long. Valves well developed and surface sculpturing varying from almost smooth to highly granular. The head valve bears an insertion plate which lacks notches, the intermediate valves lack insertion plates whilst the tail valve may or may not have an insertion plate. The articulamentum produces well developed, robust sutural laminae on valves II–VIII. The girdle is narrow and is covered by spicules and granules sometimes interspersed with longer spines. The girdle edge has a fringe of densely packed spines. Members of this family are all merobranchial. The family contains two genera, only one of which is represented in British waters.

Genus HANLEYA Gray, 1857

Contains species of varying sizes. The valves are distinctly sculptured with randomly distributed, raised dorsal papillae. The head valve bears an insertion plate lacking any notches, valves II–VIII have no insertion plates. The girdle is narrow and densely covered by small spicules with numerous larger spines randomly distributed.

Type species: *Hanleya debilis* Gray, 1857 = *Chiton hanleyi* Bean in Thorpe, 1844.

Hanleya hanleyi (Bean in Thorpe, 1884)
(Fig. 16)

Chiton hanleyi Bean in Thorpe, 1844
Hanleya debilis Gray, 1857
Hanleya hanleyi Dautzenberg and Fischer, 1912
Lepidopleurus carinatus Dall, 1927
Hanleya dalli Kaas, 1957
Laminoplax dalli Ferreira, 1981

A merobranch chiton with an oblong shape (breadth : length ratio about 0.5) and with a recorded maximum length of 22 mm. The girdle is narrow, the shell width : total width ratio being about 0.8 (Fig. 16A).

The shell valves are robust, without notches and have an insertion plate on the head valve only. Valves II–VIII have relatively small, rounded-triangular sutural laminae (Fig. 16F, G). The head valve is disproportionately large while the intermediate valves are moderately tall, keeled and with a very pronounced beak (Fig. 16F). The tail valve is rounded with a prominent mucro centrally located (Fig. 16G).

The dorsal surface of the valves is very coarsely sculptured when seen under a hand-lens; the jugal, median and lateral areas of the intermediate

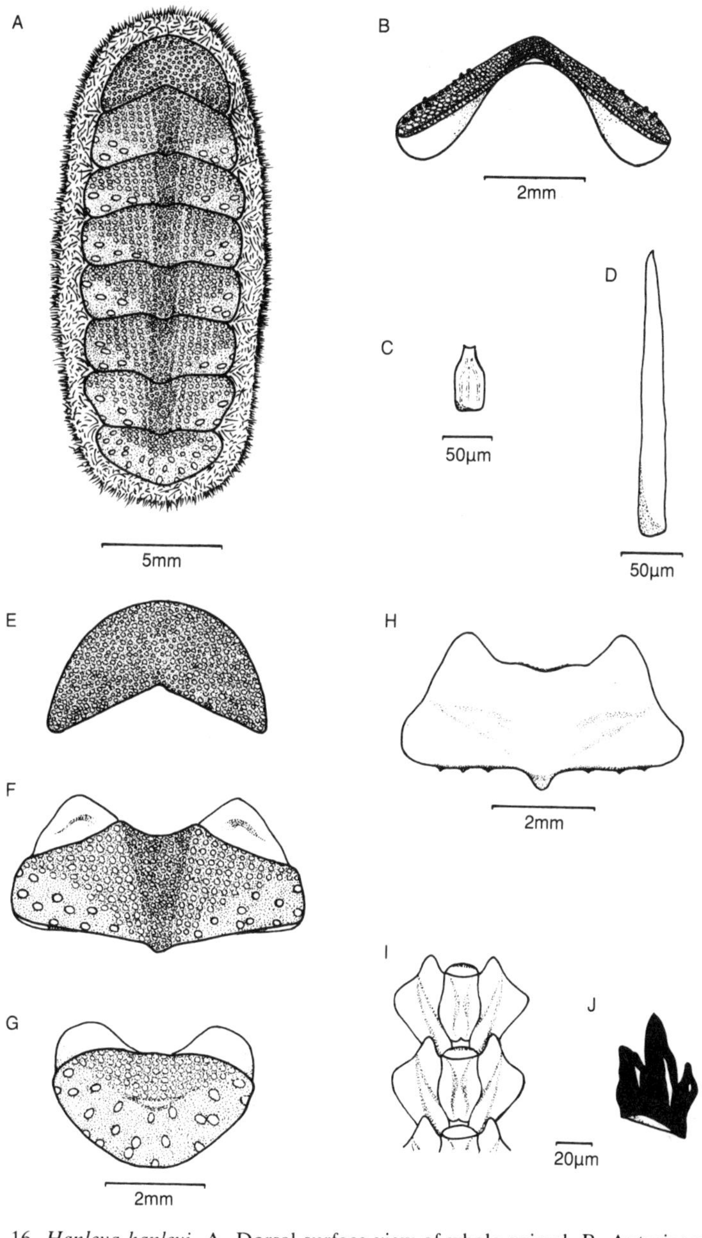

Fig. 16. *Hanleya hanleyi*. A. Dorsal surface view of whole animal. B. Anterior view of intermediate valve. C. Girdle spicule. D. Girdle spine. E. Dorsal view of head valve. F. Dorsal view of intermediate valve. G. Dorsal view of tail valve. H. Ventral view of intermediate valve. I. Median and intermediate radula plates. J. Uncinate blade of radula.

valves are clearly defined due to differences in the arrangement, density and size of the dorsal papillae (Fig. 16A, F). Examination under the dissecting microscope reveals that the jugal area is covered by randomly distributed, closely packed small papillae while the adjacent median areas have precise rows of larger papillae running antero-posteriorly producing a furrowed appearance (Fig. 16F). The lateral area has relatively few randomly scattered large papillae. Each papilla bears an apical cap positioned centrally with a crescent-shaped ring of subsidiary caps around it. The valve surface is generally too rough to allow reliable identification of growth interruption lines. The ventral surface of all the valves is unperforated, this feature being characteristic of all members of the Order Lepidopleurina. Colouration is naturally white or beige but darker, blackened specimens may be found due to the deposition of iron and manganese salts on the valve surface.

The girdle is covered by spicules (50 μm long) each with two or three longitudinal ridges (Fig. 16C); long, smooth spines (200 μm long) (Fig. 16D) are randomly distributed over the girdle surface, however, they appear to be more abundant in the areas adjacent to the intersections of the intermediate valves and around the margin of the girdle.

The median radular plate is longer than it is wide, with rounded angles and a small blade at the distal end; there is also a characteristic notch on its posterior margin (Fig. 16I). The intermediate plates are broad and wing-like, extending just beyond the distal end of the median plate. The uncinate blade has three denticles of different sizes (Fig. 16J).

H. hanleyi is relatively rare occurring sublittorally between the depths of 15–550 m, but most commonly between 100 m and 300 m. It is found chiefly on sandy bottoms attached to shell debris or small pebbles. It is an Atlantic-Boreal species and has been recorded from the Arctic Ocean, Barents Sea, off the coast of Scandinavia, Britain and from the Mediterranean.

Order ISCHNOCHITONINA Bergenhayn, 1930

Animals range in size from small to quite large. The head valve is undifferentiated with an anterior area only; the intermediate valves are differentiated with a jugal area, lateral and median triangles all defined; the tail valve has lateral and posterior areas differentiated. The articulamentum is well-developed producing insertion plates bearing notches of varying numbers on all valves, valves II–VIII have prominent sutural laminae. The girdle varies in relative width and in ornamentation and does not encroach to any degree over the valves.

Family ISCHNOCHITONIDAE Dall, 1889

The body is oval to elongate, rarely exceeding 60 mm in length. Valve dorsal surface sculpturing is variable but generally costate and reticulate, and divides the intermediate valves into lateral and central regions as a result of different orientations of the papillae. All valves have insertion plates, the head and tail valves have numerous notches, the intermediate valves have between one and three pairs of notches depending on the species. The insertion plates are not pectinated. The articulamentum produces well-developed sutural laminae on valves II–VIII. The girdle is covered by scales, spicules and minute hairs and is fringed by bluntly pointed spines. This family contains seven genera, four of which have representatives in the British fauna.

Genus LEPIDOCHITONA Gray, 1821

Oval to elongate-oval animals with valve sculpturing varying from almost smooth to uniformly granulose with the lateral areas only very indistinctly defined. The head and tail valves each have many notches in their insertion plates, the intermediate valves typically have a single pair of notches. The girdle is covered in granules of different sizes and fringed by a row of bluntly pointed spines. Animals range from merobranchial to holobranchial.

Type species: *Chiton marginatus* Pennant, 1777 = *Lepidochitona cinerea* (Linnaeus, 1767).

Lepidochitona cinereus (Linnaeus, 1767)
(Fig. 17)

Chiton marginatus; Brown, 1823
Chiton cinereus; Hanley, 1855
Craspedochilus marginatus Brogger, 1901
Trachydermon cinereus Thiele, 1902
Ischnochiton marginatus Hildalgo, 1917
Lepidochitona cinereus Winckworth, 1932
Lepidochiton cinereus; Van Benthem Jutting, 1936
Lepidochitona cinerea; Laghi, 1977

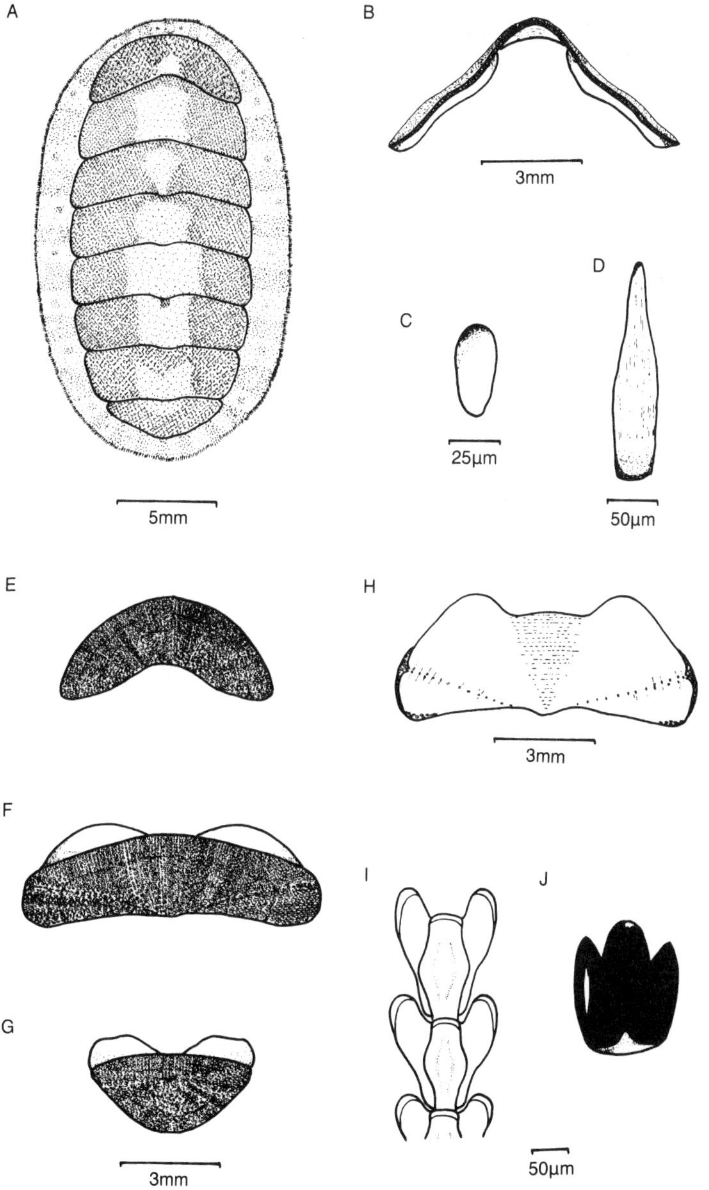

Fig. 17. *Lepidochitona cinereus*. A. Dorsal surface view of whole animal. B. Anterior view of intermediate valve. C. Girdle granule. D. Girdle marginal spine. E. Dorsal view of head valve. F. Dorsal view of intermediate valve. G. Dorsal view of tail valve. H. Ventral view of intermediate valve. I. Median and intermediate radula plates. J. Uncinate blade of radula.

An almost holobranch chiton with a moderately tall, oval shape (breadth : length ratio 0.6–0.7) and with a recorded maximum length of 28 mm. The girdle is narrow with a shell width : total width ratio of about 0.75 (Fig. 17A).

The shell valves are strong with insertion plates bearing notches. The head and tail valves typically have 8 and 10–12 notches respectively but these numbers are extremely variable with ranges of 7–11 and 6–16 notches having been recorded (Baxter, 1982); the intermediate valves (II–VII) have a single pair of notches (Fig. 17H). Valves II–VIII have relatively small, rounded, flat sutural laminae (Fig. 17F, G). The intermediate valves are moderately tall with a definite keel (Fig. 17B) and only a slight beak (Fig. 17F). The posterior valve is moderate in size, roughly triangular with a distinct, almost central mucro (Fig. 17G).

The dorsal surface of the valves is granular when seen under a hand-lens and the intermediate valves have lateral areas only indistinctly defined by the different orientation of the granules (Fig. 17F). Examination under the dissecting microscope reveals the granular sculpturing to be due to alternating rows of slightly raised dorsal papillae; each dorsal papilla bears an apical cap situated centrally with a large number of subsidiary caps distributed randomly around it, these being confined mainly to the raised papilla. Growth interruption lines are sometimes visible. The ventral surface of all valves is perforated by notch rays, and on valves II–VIII there are large numbers of holes arranged in a regular pattern perforating the jugal area (Fig. 17H). A very few holes are also present along the posterior edge of the intermediate valves. Colouration is extremely variable and it cannot be said that there is a 'typical' colour; it includes various shades of white, orange, red, blue, green, brown and black with many of the animals being ornately patterned with more than one colour. Although there do appear to be repeated patterns, individuals within a population vary enormously; the pattern expressed on any one valve appears to be independent of that expressed on adjacent valves.

The girdle is covered by densely packed, rounded granules of varying size (25–50 μm diameter), (Fig. 17C) and bordered by a fringe of gold coloured spines (150–200 μm long) each having 5–7 longitudinal grooves (Fig. 17D). The girdle is typically banded with alternating colours (Fig. 17A).

The median radular plate (Fig. 17I) is elongate with a central rib and a pronounced blade at its distal end; the intermediate plates are elongate, curved and wing-like and extend up to one quarter of their length beyond the anterior end of the median plate. The uncinate blade is crown-shaped with a central cusp and two lateral denticles of similar size and has a characteristic notch at its point of attachment to the plate (Fig. 17J).

L. cinereus is common throughout the littoral zone and extends into the *Laminaria* zone. It is usually found on stony/rocky bottoms, often on the underside of stones lightly embedded in sand and gravel in rock pools. It has a range extending from the Lofoten Islands in the north, through Scandinavia, western Europe and to the north coast of the Mediterranean. It is abundant around the entire coastline of the British Isles.

Genus TONICELLA Carpenter, 1873

Oval animals of medium size, valve sculpturing varying from almost smooth to micro-granular with the lateral areas only slightly raised and weakly defined. Typically coloured in bright reds, pinks and oranges often with a mottling of other colours. The head and tail valves have many notches in their insertion plates, the intermediate valves typically have a single pair of notches. Valves II–VIII have strong, rounded sutural laminae. The girdle appearance ranges from leathery to finely granular, and is covered by different sized granules and bordered by a row of spines. All the members of this genus are merobranchial. Two species occur in British waters.

Type species: *Chiton marmoreus* Fabricius, 1780.

Tonicella rubra (Linnaeus, 1767)
(Fig. 18)

Chiton ruber Linnaeus, 1676
Tonicia rubra (Linnaeus) Gray, 1847
Leptochiton ruber (Linnaeus) Adams and Adams, 1858
Trachydermon ruber (Linnaeus) Carpenter, 1873
Boreochiton ruber ('Lowe' [=Linnaeus]) Sars, 1878
Ischnochiton ruber (Linnaeus) Pilsbry, 1892
Tonicella rubra (Linnaeus) Sykes, 1893
Tonicella ruber index Balch, 1906
Lepidochiton ruber (Linnaeus) La Rocque, 1953

A merobranch chiton with an oval shape (breadth : length ratio of about 0.6) and with a recorded maximum length of 21 mm. The girdle is relatively narrow with a shell width : total width ratio of 0.7–0.8 (Fig. 18A).

The shell valves are strong, with insertion plates bearing notches on all valves. The head and tail valves typically have 8 and 9–10 notches respectively but these numbers can vary considerably; the intermediate valves (II–VII) each have a single pair of notches (Fig. 18H). Valves II–VIII have markedly curved, prominent sutural laminae (Fig. 18F). The intermediate valves are rounded and moderately keeled (Fig. 18B) and with a distinct beak (Fig. 18F). The posterior valve is roughly triangular in shape with a slightly anteriorly displaced prominent mucro (Fig. 18G).

The dorsal surfaces of all valves are smooth when seen under a hand lens, apparently without raised dorsal papillae; the intermediate valves have large but only very indistinctly defined lateral areas (Fig. 18F). Although without obvious raised papillae, examination under the dissecting microscope reveals a stippled valve surface due to large numbers of apical caps, each surrounded by an identifiable, distinct group of associated subsidiary caps. There are often very distinct growth disturbance rings on the valves (Fig. 18A). The ventral surface of all valves is perforated by notch rays and on valves II–VIII

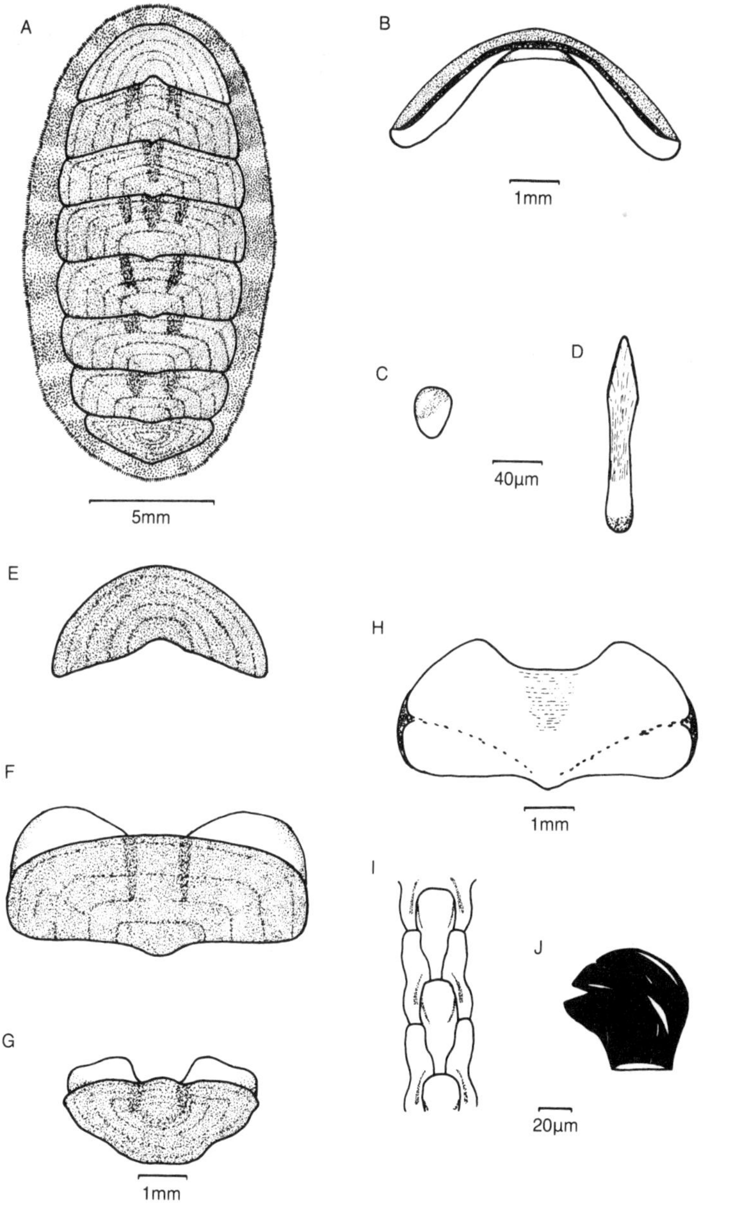

Fig. 18. *Tonicella rubra*. A. Dorsal surface view of whole animal. B. Anterior view of intermediate valve. C. Girdle granule. D. Girdle marginal spine. E. Dorsal view of head valve. F. Dorsal view of intermediate valve. G. Dorsal view of tail valve. H. Ventral view of intermediate valve. I. Median and intermediate radula plates. J. Uncinate blade of radula.

there are also large numbers of holes arranged in a regular pattern penetrating the jugal area (Fig. 18H). A few holes are also present on the ventral surface of the intermediate valves running along their posterior edge. Colouration is invariably between pink and brick red with patches and/or streaks of cream and light brown. There appears to be some continuity in the patterning on the intermediate valves; this takes the form of either a general overall mottling, light or dark streaks running antero-posteriorly on either side of the jugal area, or light or dark wedge-shaped patches on the keel extending backwards to the beak (Fig. 18A).

The girdle is covered by closely packed oval granules (long diameter – 40 μm) (Fig. 18C) and the edge has a fringe of golden coloured spines (140 μm long) each having 8–12 fine, longitudinal grooves (Fig. 18D). The girdle is typically banded with alternating cream and orange colouration (Fig. 18A).

The median radular plate has a widened anterior end and bears a distinct blade; the intermediate plates are broadly wing-like with a deep groove on the outer margin and are longer than the median plate (Fig. 18I). The uncinate blade has three distinct denticles of different sizes and is as long as it is wide (Fig. 18J).

T. rubra can be found from around low water down to depths of 300 m, but is most common between 10 m and 40 m although never occurring in any great density. It shows a preference for stony bottoms but also occurs on mixed sand and pebble. It is a typical North Atlantic-Boreal species and has been recorded in the Barents and White Seas, occasionally off Greenland and is common along the coasts of Norway, Iceland, the Faroes and the whole of the British Isles.

Tonicella marmorea (Fabricius, 1780)
(Fig. 19)

Chiton marmoreus Fabricius, 1780
Tonicia marmorea (Fabricius) Adams and Adams, 1854
Tonicella marmorea (Fabricius) Carpenter, 1873
Boreochiton marmoreus (Fabricius) Sars, 1878
Tonicella submarmorea (Middendorff) Dall, 1886
Tonicella marmorea Thiele, 1893
Lepidochitona marmorea (Fabricius) Dall, 1921
Lepidochiton (Tonicella) marmorea (Fabricius) La Rocque, 1953

A merobranch, although sometimes partially holobranch, chiton with an oval shape, the breadth : length ratio being about 0.6 (Fig. 19A); it is the largest of the species found in Scotland, with a maximum recorded length of 45 mm. The girdle is narrow with a shell width : total width ratio of about 0.7 (Fig. 19A).

The valves are strong with insertion plates bearing notches. The head and tail valves typically have 8–10 and 7–9 notches respectively, but numbers vary considerably outside these limits with recorded ranges of 7–13 and 6–11 respectively, while the intermediate valves (II–VII) each have one pair of notches (Baxter & Jones, 1986). Valves II–VIII bear wide, flattened sutural laminae (Fig. 19F, G). The intermediate valves are moderately tall with a pronounced keel (Fig. 19B) but have only an indistinct, often almost absent beak (Fig. 19F). The posterior valve is relatively small and roughly elliptical in shape bearing an anteriorly displaced mucro (Fig. 19G).

The dorsal surface of the shell valves appears smooth to the naked eye but is actually covered by numerous minute granules, revealed when seen under a hand-lens; the various regions of the intermediate valves are poorly differentiated, the lateral area only being defined by a slightly raised diagonal ridge (Fig. 19F). Examination under the dissecting microscope shows the valve surface to be covered by very small, only slightly raised, dorsal papillae arranged in a regular, uncrowded pattern; an apical cap is located on each papilla and is surrounded by its associated subsidiary caps. Distinct growth disturbance lines are often present on all the valves. The ventral surface of all valves is perforated by notch rays (Fig. 19H) and on valves II–VIII there are large numbers of holes arranged in a regular pattern penetrating the jugal area. Valve colouration in the majority of specimens varies from light red to dark brick red with irregular mottling of brown and cream, but other colour morphs such as beige, green and blue have occasionally been recorded. Adjacent valves on a single animal may have contrasting colours, the most common pattern being lightly coloured valves, I, III, V, VI and VIII with the remainder being darkly coloured. The valves of *T. marmorea* are occasionally encrusted with macro-epibiota or have deposits of manganese forming a black surface layer.

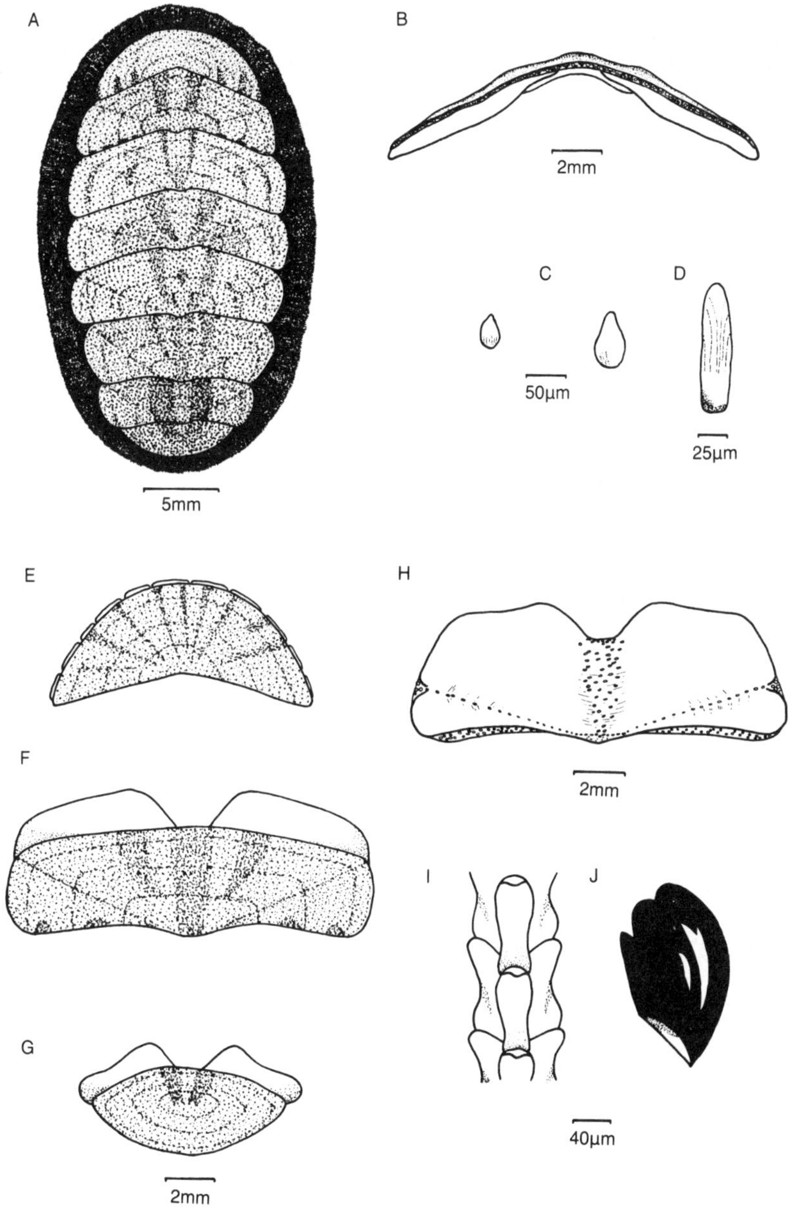

Fig. 19. *Tonicella marmorea*. A. Dorsal surface view of whole animal. B. Anterior view of intermediate valve. C. Girdle granule. D. Girdle marginal spine. E. Dorsal view of head valve. F. Dorsal view of intermediate valve. G. Dorsal view of tail valve. H. Ventral view of intermediate valve. I. Median and intermediate radula plates. J. Uncinate blade of radula.

The girdle is covered by very sparsely distributed minute granules 25–30 µm in diameter (Fig. 19C) producing a smooth and leathery appearance. The edge of the girdle is fringed by a ring of spatulate spines (length 100 µm) with longitudinal grooves (Fig. 19D). The girdle colouration is variable ranging from deep purplish red to green and may be banded with orange.

The median radular plate is slightly rounded anteriorly with an indistinct blade (Fig. 19I); the intermediate plates are narrow, wing-like with a shallow incision at the distal end and extend approximately one quarter of their length beyond the anterior end of the respective median plate. The uncinate blade is narrow, approximately twice as long as it is wide with three denticles, the outer and the mid-denticles being separated by a deep notch (Fig. 19J).

T. marmorea occurs from low water to around 200 m depth on a variety of substrates, but prefers bedrock lying at a variety of angles or a stone and pebble bottom. It is an Atlantic-Boreal species, one of the few wide ranging *Tonicella* species, occurring in the Arctic Ocean, off the coast of Norway and Sweden, Britain, Greenland, and the Atlantic coast of North America. In Britain it is restricted to the north, extending southwards only as far as Northumberland on the east and North Wales on the west. It occurs in large numbers in many of the sea-lochs of Scotland where densities of 40–50 per square metre are not uncommon and it is an important grazer within the ecosystem.

Genus ISCHNOCHITON Gray, 1847

Relatively small, oval animals which reach a maximum size of 50 mm. Valves are sculptured with raised dorsal papillae, the lateral triangles are often clearly defined by being slightly raised. The head valve is disproportionately large and undifferentiated, the tail valve bears a well developed mucro. All valves have insertion plates, those on the head and tail valves have numerous notches, those on the intermediate valves typically have a single pair of notches. The girdle is densely covered with relatively large, evenly distributed scales.

Type species: *Chiton textilis* Gray, 1828.

Ischnochiton albus (Linnaeus, 1767)
(Fig. 20)

?*Chiton minimus* Gmelin, 1791
?*Chiton oryza* Spengler, 1797
Chiton aselloides Lowe, 1825
Chiton sagrinatus Couthouy, 1838
Lophyrus albus (Linnaeus, 1767) var. *infuscatus* Sparre Schneider, 1881

A merobranch chiton with a slightly elongated oval shape (breadth : length ratio of 0.6–0.7) and a recorded maximum length of 14 mm. The girdle is narrow with a shell width : total width ratio of about 0.8 (Fig. 20A).

The shell valves are thin and fragile with insertion plates bearing notches. The head and tail valves typically have 13 and 11–12 notches respectively, but these figures are variable; the intermediate valves (II–VII) have a single pair of notches. Valves II–VIII have small, rounded, flat sutural laminae which taper inwards merging into the jugal sinus (Fig. 20F, G). The intermediate valves are of medium height with a keel (Fig. 20B), and a slight beak (Fig. 20F). The tail valve is moderately large, rounded and with a prominent anteriorly placed mucro (Fig. 20G).

The dorsal surface of the valves is only slightly granular when seen under a hand lens and the intermediate valves have lateral areas very weakly defined by being only slightly raised (Fig. 20F). Examination under the dissecting microscope reveals the granular appearance due to alternating rows of slightly raised oval dorsal papillae. Growth interruption lines are often visible (Fig. 20A). The ventral surface of all valves is perforated by notch rays and on valves II–VIII there are large numbers of holes arranged in a regular pattern perforating the jugal area (Fig. 20H). A number of holes running along the posterior edge of the intermediate valves are also present. Colouration is typically off-white or cream, although the animals may often be partly covered by black deposits of iron or manganese salts.

The girdle is most distinctive, being covered by very large, oval scales

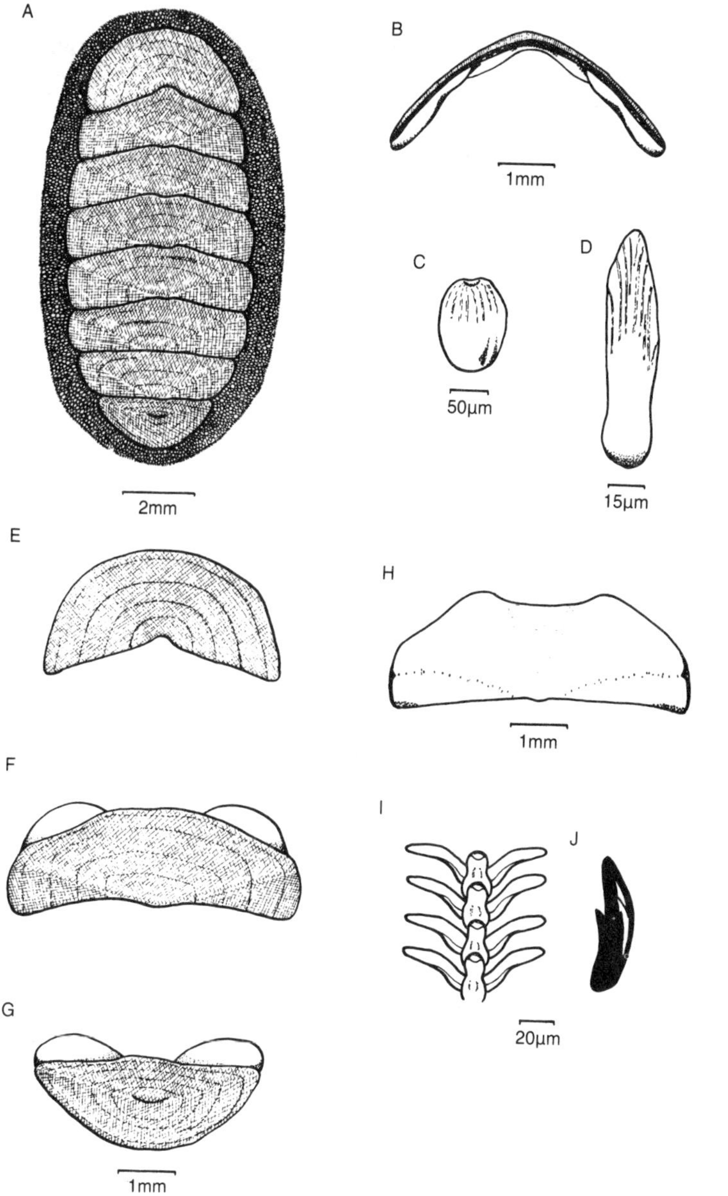

Fig. 20. *Ischnochiton albus*. A. Dorsal surface view of whole animal. B. Anterior view of intermediate valve. C. Girdle scale. D. Girdle marginal spine. E. Dorsal view of head valve. F. Dorsal view of intermediate valve. G. Dorsal view of tail valve. H. Ventral view of intermediate valve. I. Median and intermediate radula plates. J. Uncinate blade of radula.

(100 μm diameter) (Fig. 20C), and bordered by a fringe of spatulate, grooved spines (75 μm long) (Fig. 20D). The girdle is typically unbanded and appears much more granular than the valve surface when examined under a dissecting microscope. It may also be blackened by deposits of iron or manganese.

The median radular plate is wide posteriorly with a narrow neck-like anterior portion and an obvious blade; the intermediate plates are wing-like extending forward past the anterior tip of the median plate (Fig. 20I). The uncinate blade is long and narrow with two denticles, one very much larger than the other (Fig. 20J).

I. albus is uncommon but has been recorded from around most of the British Isles. It occurs on a wide variety of bottoms extending from around low water to depths exceeding 500 m, but is most common on stone and pebble substrates down to about 100 m. It is generally distributed throughout the Boreal and Arctic Seas occurring off the coasts of Greenland, Scandinavia, North America and Britain extending into the White, Barents and Baltic Seas.

Genus CALLOCHITON Gray, 1847

Relatively small, oval chitons, typically reddish in colour although somewhat variable. Valves appear almost smooth with only very slightly raised dorsal papillae, resulting in the lateral areas being only very indistinctly defined. The head and tail valves bear insertion plates with numerous notches each, the intermediate valves typically have three pairs of notches. The sutural laminae on valves II–VIII are prominent and continuous across the jugal area. The girdle is covered in densely packed, definitely arranged, needle-like spicules and has a fringe of short spines.

Type species: *Chiton laevis* Montagu, 1803 (not Pennant, 1777) (=*Chiton septemvalvis* Montagu, 1803).

Callochiton achatinus (Brown, 1823)
(Fig. 21)

Chiton laevis Montagu, 1803
Chiton septemvalvis Montagu, 1803
Chiton discors Maton and Rackett, 1807
Chiton achatinus Brown, 1823
Chiton euplaea Costa, 1829
Chiton doriae Capellini, 1859
Callochiton laevis Pilsbry, 1892
Callochiton doriae Thiele, 1902
Callochiton achatinus Winkworth, 1932

A holobranch chiton with a flattened, oval shape (breadth : length ratio about 0.6) and with a recorded maximum length of 32 mm. The girdle is broad with a shell width : total width ratio of about 0.6 (Fig. 21A).

The shell valves are strong with insertion plates bearing notches. The head and tail valves usually have 15–20 notches each, while the intermediate valves (II–VII) have three pairs of notches (Fig. 21H); this is the only British chiton typically to have more than one pair of notches. Valves II–VIII have slightly curved, relatively small sutural laminae which are connected across the jugal area (Fig. 21F, G). The intermediate valves are moderately curved, keeled (Fig. 21B) and with a pronounced beak (Fig. 21F); the posterior valve is relatively large and rounded with an indistinct, anteriorly displaced mucro (Fig. 21G).

The dorsal surface of the valves is slightly granular when seen under a hand lens; the intermediate valves have clearly defined lateral areas (Fig. 21A, F). Examination under the dissecting microscope reveals that the shell surface is very finely grooved with only very slightly raised dorsal papillae around each apical cap. The fine grooves on the valve surface are the result of lines of subsidiary caps running between, but not obviously associated

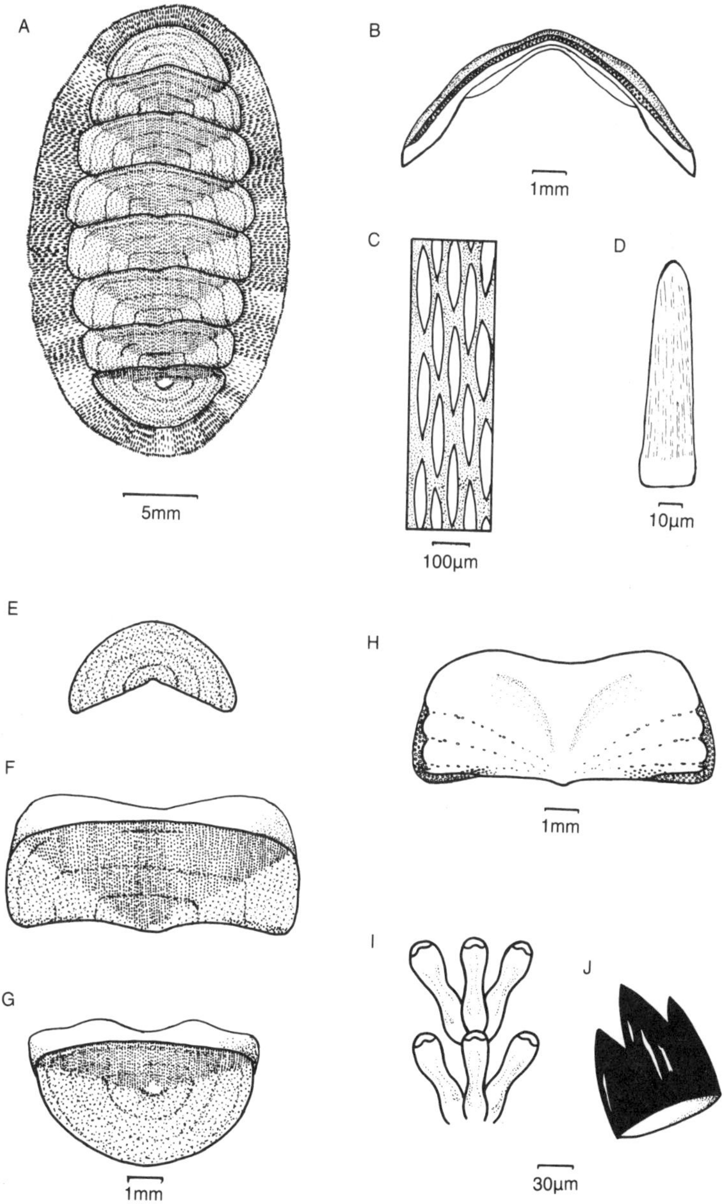

Fig. 21. *Callochiton achatinus*. A. Dorsal surface view of whole animal. B. Anterior view of intermediate valve. C. Girdle spicule. D. Girdle marginal spine. E. Dorsal view of head valve. F. Dorsal view of intermediate valve. G. Dorsal view of tail valve. H. Ventral view of intermediate valve. I. Median and intermediate radula plates. J. Uncinate blade of radula.

with, individual apical caps. The surfaces of the head and tail valves, and the lateral areas of the intermediate valves, have obvious black pigmented spots lying immediately proximal to the apical caps and this is diagnostic of *C. achatinus* within the British fauna. Under the SEM the true complexity of the valve sculpturing is revealed (Baxter & Jones, 1984). Growth interruption lines are often faintly visible. The ventral surface of all valves is perforated by notch rays; the posterior regions of the ventral surface of valves II–VII are also perforated by a large number of additional holes (Fig. 21H). Colouration is typically reddish-orange with irregular cream or brown marbling, but other colour morphs such as yellow and green are not unknown.

The girdle is covered by densely packed acicular spicules (length 200 μm) producing a snake-skin like appearance with diamond patterning which is characteristic of this species (Fig. 21C). The girdle is fringed by short (100 μm maximum length), colourless spines having 6–8 longitudinal grooves (Fig. 21D). The girdle is uniformly pink in colour with bands of cream sometimes occurring, particularly at the junctions between valves I and II and valves VII and VIII (Fig. 21A).

The median radular plate is long and narrow, tapering in the middle and bearing a small blade at the anterior end; the intermediate plates are long and wing-like and also bear a small blade (Fig. 21I). The uncinate blade is crown-shaped with a large central cusp and two smaller lateral denticles of similar size (Fig. 21J).

C. achatinus is distributed around the British Isles in relatively low numbers occurring between the *Laminaria* zone and a depth of 160 m. It is most commonly found attached to the underside of stones or pebbles which are partially embedded in clean sand at about ELWS. It has a range extending from Norway in the North to the Mediterranean in the South.

Order ACANTHOCHITONINA Bergenhayn, 1930

Animals range in size from small to the largest known chitons. The valves are typically partially or completely covered by the girdle, resulting in the exposed dorsal surface of the tegmentum being reduced in area relative to the embedded articulamentum. The head valve is undifferentiated, the intermediate valves have a distinct jugal area flanked by undifferentiated latero-pleural areas, the tail valve is reduced. The girdle is smooth or spiny but not scaly, often bearing tufts of bristles.

Family ACANTHOCHITONIDAE Pilsbry, 1893

The animals in the various genera of this family are superficially quite different, but related by various valve and girdle characteristics. The girdle encroaches over the valves either partially or completely. The tegmentum of the valves is much reduced relative to the articulamentum which produces very prominent sutural laminae. The intermediate and tail valves have a well-defined jugal area flanked by uniformly sculptured latero-pleural areas, the tail valve has a prominent mucro with a relatively large posterior area. The notch numbers in the insertion plates of the various valves are variable. The girdle is relatively broad and may be either smooth or spiny, there are in addition often bundles or tufts of bristles distributed around the girdle. The Acanthochitonidae comprises five genera, only one, *Acanthochitona*, is represented in British waters.

Genus ACANTHOCHITONA Gray, 1821

Small to medium sized animals, rather elongated with a relatively broad girdle which encroaches to a greater or lesser extent between the valves. Valve sculpturing is coarse, consisting of raised papillae of round to pear-drop shape, often arranged in radiating rows. The head valve has five notches in its insertion plate; valves II–VIII each have a single pair of notches. The girdle varies from nude to densely spiny, but invariably there is a series (typically 18) of tufts of bristles around the head valve and situated at the sutures of each valve.

The general characteristics of the members of the Genus *Acanthochitona* allow them to be easily distinguished from the members of all the other British genera. The presence of usually 18 tufts of bristles on a broad, fleshy girdle surrounding reduced valves is the most obvious diagnostic character.

The production of reliable and accurate descriptions of the various different species of this genus is difficult due to a confusing array of synonyms and perseverance with some basically inadequate taxonomic characters. Kaas (1985) has reviewed thoroughly the history of the nomenclature of the genus and clarified the status of the various common synonyms. It is in essence this revised nomenclature which is adopted here. It must be noted, however, that the species descriptions that exist in the literature were derived from either single specimens or very small numbers of individuals and as such are not wholly reliable in representing any intraspecific variability.

Many of the characteristics used to describe the different species are both variable and dependent, to a certain extent, on the experience of the person involved and in effect require at the very least comparative specimens to be available. Three species are currently recognised as present in the British fauna, but considerable doubt must be cast over two of these as there is an increasing weight of evidence to support the idea that they constitute a single 'variable' species. The three which are currently recognised are: *Acanthochitona fascicularis*, *Acanthochitona crinitus* and *Acanthochitona discrepans*. These are defined by the following descriptions:

Acanthochitona fascicularis. The valve dorsal surface is finely granular with small, round papillae very densely distributed in a regular pattern. The girdle is very broad and densely covered by fine, slightly curved spicules with longer, thicker spicules interspersed amongst them.

Acanthochitona crinitus. The valve dorsal surface is coarsely granular with large pear-shaped papillae irregularly distributed. The girdle is broad, covered by large numbers of spicules with randomly distributed spines which tend to be concentrated towards the outer half of the girdle.

Acanthochitona discrepans. The valve dorsal surface is granular with oval-shaped papillae arranged in a more or less regular pattern. The girdle is broad and densely covered by spicules.

Type species: *Chiton fascicularis* Linnaeus, 1767.

Acanthochitona fascicularis (Linnaeus, 1767)
(Fig. 22)

Chiton fascicularis Linnaeus, 1767
Acanthochites communis Risso, 1826
Chiton fascicularis var. *major* Philippi, 1836
Chiton crinitus; Sowerby, 1840
Chiton discrepans; Sowerby, 1840
Acanthochites carinatus H. Adams and Angas, 1864
Acanthochites hamatus de Rochbrune, 1882
Acanthochiton(a) communis; Winckworth, 1926
Acanthochiton discrepans var. *angustivalvus* Bergenhayn, 1931
Acanthochiton communis forma *barashi* Leloup, 1969
Acanthochitona bonairensis Kaas, 1972

A merobranch chiton with a rather elevated oblong shape (breadth : length ratio of 0.6–0.7) and with a recorded maximum length of 60 mm, the largest British chiton. The girdle is very broad with a shell width : total width ratio of 0.3–0.4 (Fig. 22A).

The shell valves are heavy and strong with large insertion plates bearing notches on all valves (Fig. 22F, G). The head valve typically has five notches (Fig. 22F) while the intermediate valves and tail valve each have a single pair of notches. Valves II–VIII have rounded, very prominent sutural laminae (Fig. 22G, H). The intermediate valves are elevated, with a definite keel (Fig. 22B) and a moderate beak (Fig. 22G); the antero-lateral areas of the tegmentum are incurved, creating a shield-shaped dorsal surface (Fig. 22G). The tail valve is disproportionately small, oval and with a much reduced posterior region (Fig. 22H).

The dorsal surface of the head valve, the median and lateral areas of the intermediate valves and the posterior area of the tail valve are granular when seen under a hand-lens. The jugal area of valves II–VII and the central area of valve VIII are clearly defined as a narrow, triangular area with faint longitudinal grooves; the median and lateral areas are undifferentiated (Fig. 22G, H). Examination under the dissecting microscope reveals that the granular sculpturing is due to densely packed, small, round, raised dorsal papillae (Fig. 22G), which are arranged in precise, diverging rows. The jugal area of the intermediate valves and the central area of the tail valve are covered by antero-posterior orientated lines of merged dorsal papillae. Each dorsal papilla bears an approximately centrally situated apical cap with only a very few associated subsidiary caps. The ventral surface of valves II–VIII are perforated by holes in the jugal area only (Fig. 22I). Valve colouration is variable and often mottled with combinations of olive green and reddish-brown; the central less granular area is often pale, almost white with dark streaks thus creating a striking V-shaped region.

The girdle has 18 tufts of long colourless bristles (500–1000 μm long) (Fig. 22D); four are arranged around the head valve and the remainder are located

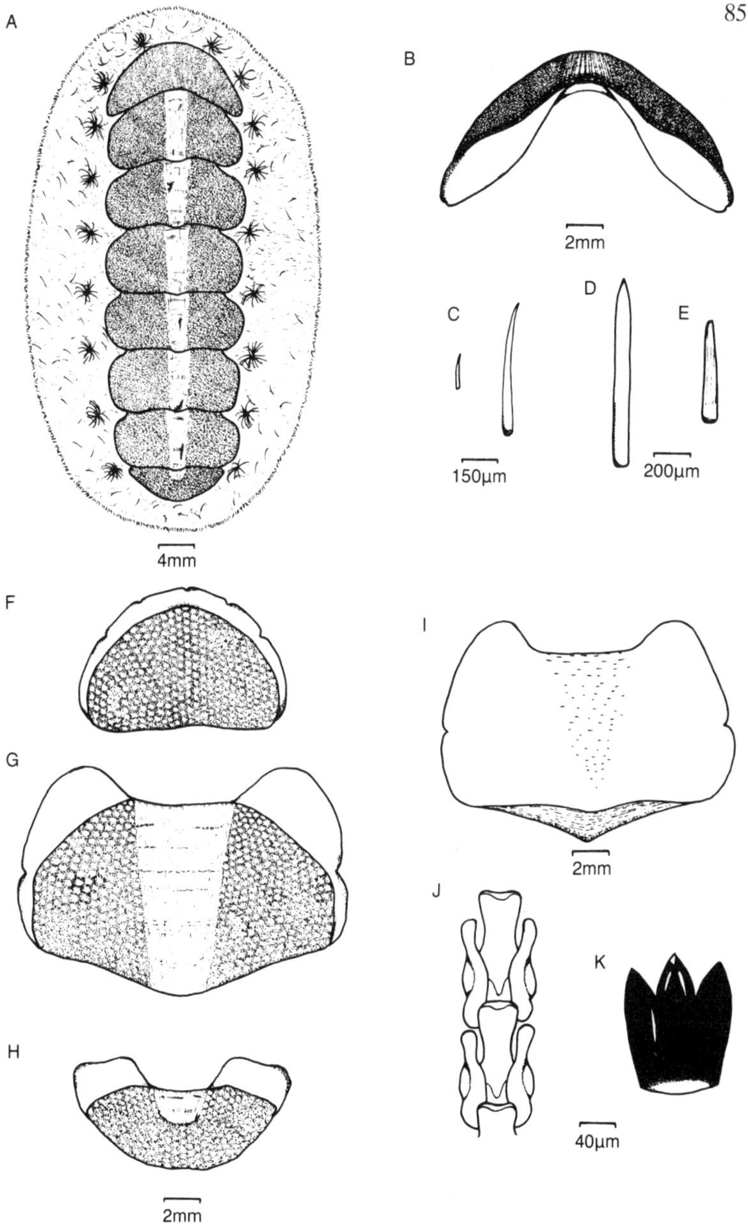

Fig. 22. *Acanthochitona fascicularis*. A. Dorsal surface view of whole animal. B. Anterior view of intermediate valve. C. Girdle spicule and spine. D. Girdle bristle. E. Girdle marginal spine. F. Dorsal view of head valve. G. Dorsal view of intermediate valve. H. Dorsal view of tail valve. I. Ventral view of intermediate valve. J. Median and intermediate radula plates. K. Uncinate blade of radula.

at each intersection of the valves (Fig. 22A). The remainder of the girdle is densely covered by short, slender, very slightly curved spicules 100–150 μm long (Fig. 22C), with larger, more robust spines about 500 μm long (Fig. 22C) scattered amongst them. The girdle margin is fringed by a dense row of spines about 500 μm long (Fig. 22E).

The median radular plate is long and narrow with a slight constriction in the middle, bearing a small blade at the anterior end and a peg-like projection at its posterior end; the intermediate blades are shorter than the central plate (Fig. 22J). The uncinate blade has three denticles, the central one being slightly larger than the other two (Fig. 22K).

A. fascicularis is found from low water to a depth of 50 m on rocks and boulders or in crevices. It is a southern species occurring along the coast of Europe from the Channel Islands into the Mediterranean and Adriatic Seas. Its distribution around the shores of the British Isles is not clearly defined, but is confined to the southern England and the Channel Islands.

Acanthochitona crinitus (Pennant, 1777)
(Fig. 23)

Chiton crinitus Pennant, 1777
Chiton onyx Spengler, 1797
Chiton fascicularis; Brown, 1827
Chiton fascicularis var. *minor* Philippi, 1836
Chiton fascicularis var. *attenuata* Jeffreys, 1865
Acanthochites (ton) adansoni de Rochbrune, 1881
Acanthochites (ton) bouvieri de Rochbrune, 1881
Acanthochitona crinitus; Winckworth, 1926
Acanthochiton fascicularis; Leloup, 1936
Acanthochiton oblongus Leloup, 1981

A merobranch chiton with a relatively flat, extended oval shape (breadth : length ratio of 0.5–0.6) and with a recorded maximum length of 34 mm. The girdle is very broad with a shell width : total width ratio of 0.4–0.5 (Fig. 23A).

The shell valves are heavy and strong with large insertion plates bearing notches on all valves. The head valve typically has five notches (Fig. 23F), while the intermediate and tail valves each have a single pair of notches. Valves II–VIII have very prominent sutural laminae (Fig. 23G, H). The intermediate valves are flat with only a slight keel (Fig. 23B) and a prominent beak (Fig. 23G); the antero-lateral areas of the tegmentum are only moderately incurved (Fig. 23G). The tail valve is disproportionately small, roughly oval in shape and with a much restricted posterior area (Fig. 23H).

The dorsal surface of the head valve, the median and lateral areas of the intermediate valves and the posterior area of the tail valve are coarsely granular when seen under a hand-lens. The jugal area of valves II–VII and the central area of valve VIII are clearly defined by their almost smooth texture with only very faint antero-posterior grooves; the median and lateral areas of the intermediate valves are undifferentiated (Fig. 23G). Examination under the dissecting microscope reveals that the granular sculpturing is due to dorsal papillae whose arrangement and shape range from being quite randomly and widely distributed, large and pear-shaped (Fig. 23G) to relatively closely and evenly distributed, medium-sized and oval. Individuals with various intermediate expressions of these characteristics have been found by the authors in populations collected from single locations. The jugal area of the intermediate valves and the central area of the tail valve are covered by closely arranged, only slightly raised lines of merged dorsal papillae, thus producing the grooved effect seen under the hand-lens. Each raised dorsal papilla bears an apical cap situated towards its anterior end surrounded by a large number (15–25) of randomly distributed subsidiary caps. The ventral surface of valves II–VIII is perforated by numerous holes in the jugal area. Valve ground colouration is typically olive green with varying degrees of any combination of brown, orange, yellow, pink or white mottling. The jugal area on the

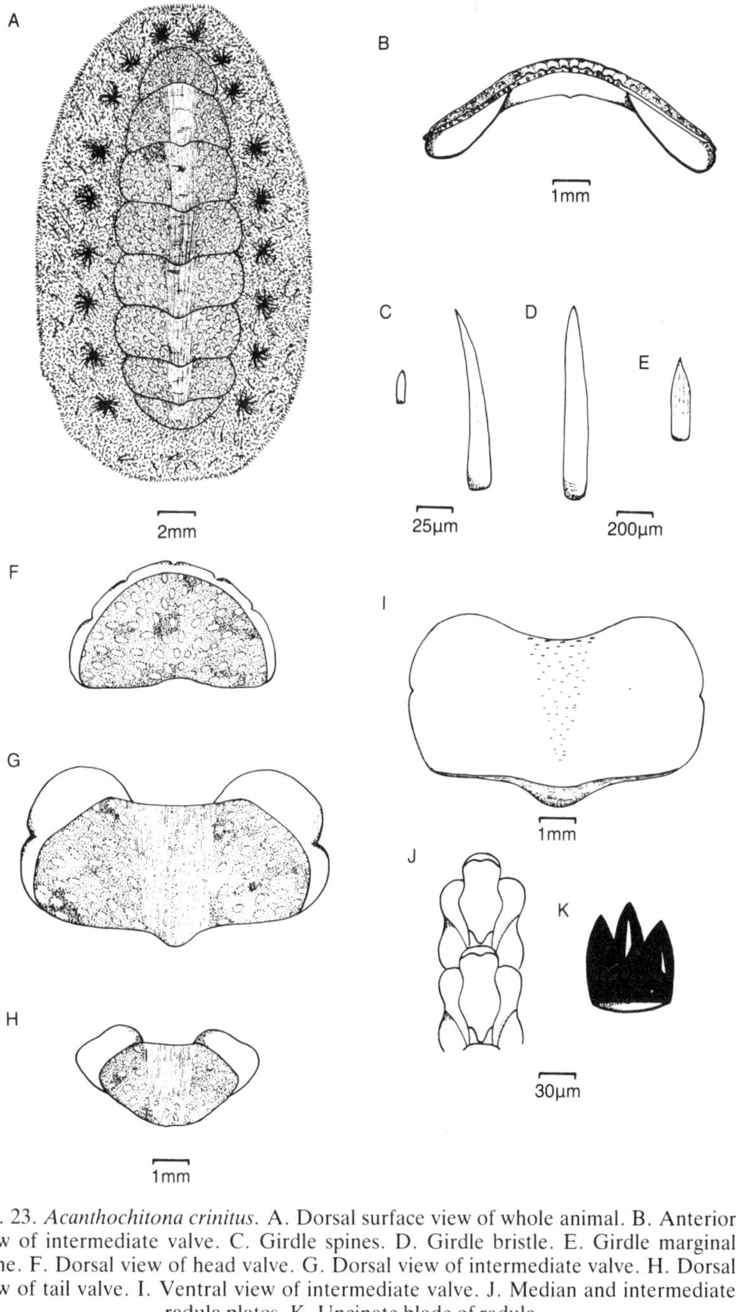

Fig. 23. *Acanthochitona crinitus*. A. Dorsal surface view of whole animal. B. Anterior view of intermediate valve. C. Girdle spines. D. Girdle bristle. E. Girdle marginal spine. F. Dorsal view of head valve. G. Dorsal view of intermediate valve. H. Dorsal view of tail valve. I. Ventral view of intermediate valve. J. Median and intermediate radula plates. K. Uncinate blade of radula.

intermediate valves and the central area on the tail valve are usually lighter in colour, either white or beige, with occasional darker streaks thus creating a striking V-shaped area. The valves are sometimes encrusted with epibiota.

The girdle has 18 tufts of long, colourless bristles (600–1000 μm long), (Fig. 23D); four are arranged around the head valve, the remainder are located at each intersection of the valves (Fig. 23A). The remainder of the girdle is very densely covered by short spines 25–30 μm long (Fig. 23C), with longer, curved spines (125–150 μm long) bearing numerous longitudinal grooves (Fig. 23C) randomly distributed to a greater or lesser degree over the surface of the girdle and tending to be more numerous towards the outer regions. The girdle margin is fringed with a dense row of spines (300–400 μm long) bearing longitudinal grooves (Fig. 23E).

The median radular plate is long and narrow with a slight constriction in the middle, bearing a small blade at the anterior end and forming a peg-like constriction at its posterior end (Fig. 23J). The intermediate plates are short and wing-like. The uncinate blade has three denticles, the middle one being slightly larger than the two outer ones (Fig. 23K).

A. crinitus is found from low water to a depth of 50 m, but is primarily an inhabitant of the low littoral and *Laminaria* zones. It lives on the underside of stones which are often lightly embedded in coarse sand or gravel. It has a range from the southwest coast of Norway along the coast of Europe to the Mediterranean and Adriatic seas. It occurs around the entire coastline of Britain and in localised areas may be present in great densities, although usually it occurs in only small numbers.

The distinguishing features which supposedly separate *A. crinitus* and *A. discrepans* are the degree of girdle spiculation and dorsal papilla density and shape. These characteristics do not allow definitive separation of these species due to the large range of intermediates which are found and, therefore, at present there is no justification to separate these species, although these findings have still to be confirmed.

SCAPHOPODA

Introduction

The Scaphopods or 'tusk-shells' form a small but distinctive class of molluscs with a fossil record extending back to the early Devonian (Palmer, 1979). Although better known than either the Caudofoveata or the Solenogastres, they have received comparatively little attention, studies of the living animal being limited mainly to the works of Yonge (1937), Dinamani (1954), Morton (1959) and Bilyard (1974). Recent studies of their classification, first by Emerson (1962) and then by Palmer (1974), have led to the scheme presented here.

The Scaphopoda appear to occupy a position intermediate between the Gastropoda and the Lamellibranchiata: the presence of a univalve shell, a buccal mass with a radula and the nature of the nervous system are characteristic of the gastropods while the digging foot, lack of cephalisation, fused mantle which is open at both ends and bilateral symmetry are typical of bivalves.

General Structure

Scaphopods are bilaterally symmetrical molluscs with a **shell** in the form of a tapering cylinder open at both ends and which is usually conspicuously curved, the concave side of the shell being dorsal. The aperture with the largest diameter is the anterior or oral end and contains both the mouth and the foot (Fig. 24A). The **body** is elongated along the postero-anterior axis and the **mantle cavity** extends over its ventral surface, opening at both ends of the shell: the **mantle** is fused along the mid-ventral line and the anterior opening can be almost entirely closed by the mantle when the animal is withdrawn into its shell. When active, both the **foot** and the **captaculae** (feeding tentacles) are protruded from the anterior end (Fig. 24B). The posterior opening is used for circulation of water, the inhalent current being created by ciliary activity while the exhalent current is produced by muscular contraction. There are no ctenidia although portions of the mantle wall are furrowed and are presumed to function as a respiratory surface.

The head forms a **proboscis** lying dorsal to the foot (Fig. 24A) but lacks eyes. The **mouth** is terminal in position with thickened, crenulated **lips** and is surrounded by a large number (up to 200 in *Antalis entalis*) of prehensile, filiform tentacles or captaculae which arise from two lobes, one on each side of the proboscis; the captaculae are highly extensible feeding organs and the tips are heavily ciliated. The **digestive system** is of the typical molluscan type and behind the proboscis there is a **buccal mass** containing a **radula** (Fig. 24C) usually formed of 18 transverse rows, each containing five teeth. After comminution by the radula, food material is passed through the **oesophagus** to the globular **stomach** which serves mainly as a gizzard; paired **digestive diverticula** open into the posterior end of the stomach. The **intestine** is looped and the **rectum** opens mid-ventrally into the mantle cavity from which the non-compacted faecal materal is extruded periodically by muscular contraction of the foot. The form of the anteriorally-placed, ventral foot divides the class into two orders. The foot is short and conical in shape in the order Dentalioida and has an **epipodial collar** (Fig. 24D), interrupted dorsally and which expands during burrowing to give the foot a three-lobed appearance. In the order Siphonodentaloida, however, the epipodial collar forms a round, crenulated **pedal disc** (Fig. 24E) which is terminal in position except in the genera *Pulsellum* and *Entalina* where a distinct median filament is present (Fig. 24E); this filament is probably homologous with the conical foot of the dentaloids.

The scaphopod shell is typically composed of four layers and its morphology is important in identification. The shell in British species varies in size from a few mm in length up to approximately 50 mm long and is tubular, tapering and open at both ends (Fig. 24A); only in the Cadulidae is the anterior opening constricted. The shell is usually curved and its sculpturing, when present, is in the form of longitudinal ridges or striations. The posterior apex of the shell may be simple, notched or lobed.

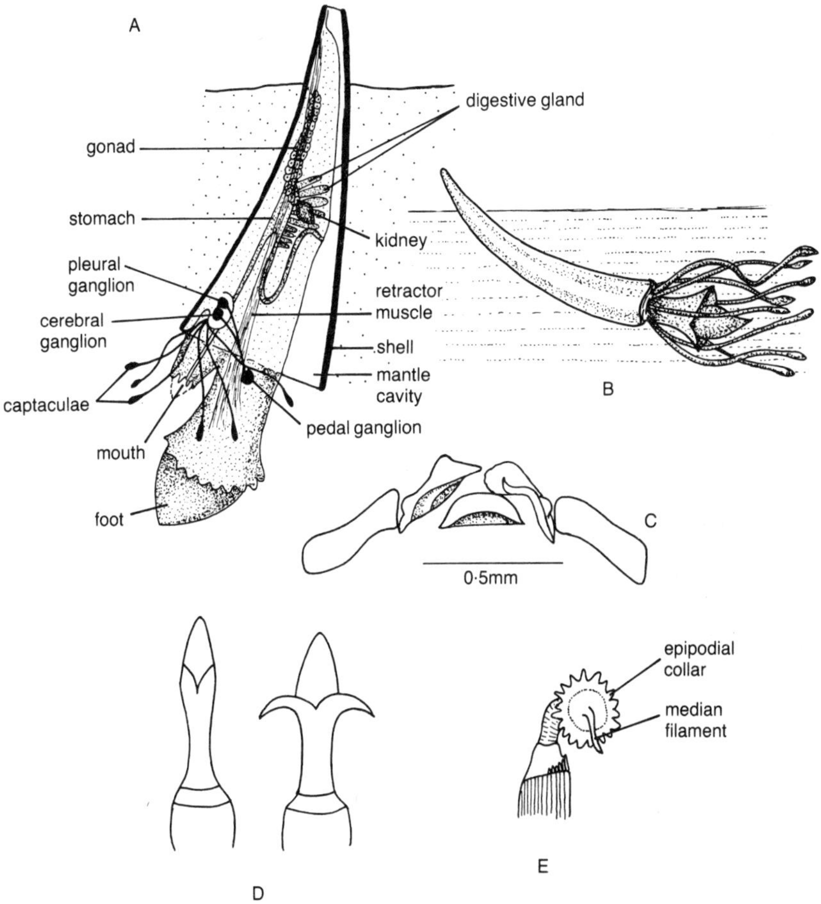

Fig. 24. General organization of scaphopods. A. General organization of a dentaloid scaphopod. B. Animal in typical feeding position with captaculae extended. C. Single, transverse row of radula teeth. D. Typical dentaloid foot: 1, in resting position; 2, in extended position used in burrowing. E. A siphonodentaloid foot with the pedal disc expanded and showing a median filament. (A. after Borradaile and Potts 1963; B after Meglitsch 1972; C–E after Palmer 1979).

The **vascular system** is extremely simple being a spacious **haemocoele** without a heart or blood vessels. There are a pair of **renal organs** which open into the mantle cavity lateral to the anus; into the right renal organ opens the duct from the single, unpaired **gonad**, the sexes being separate. The nervous system is based upon the same plan as that in the Gastropoda and the Lamellibranchiata but with an extra component, a **stomato-gastric complex** derived from the cerebral ganglion.

Biology and Ecology

Scaphopods are entirely marine and exclusively infaunal organisms living in sediments which range from muds to medium-coarse gravel. They live in water from the shallow sub-littoral to depths greater than 3000 m. In its living position (Fig. 24B) the shell is found lying with the dorsal (concave) side uppermost and the posterior tip protruding from the sediment. Studies of living animals are few and, of British species, only *Antalis (Dentalium) entalis* has been described. Usually recorded in relatively low numbers, *A. entalis* may become locally abundant (Morton, 1959).

Scaphopods are generally regarded as carnivores, feeding mainly on foraminiferans, but studies by Dinamani (1954) and Bilyard (1974) have suggested that they could be regarded as omnivores. Food is captured by the numerous captaculae and, according to Dinamani (1954), is passed back to the mouth along a ciliated tract.

Dentaloids burrow by repeated extensions and contractions of the muscular foot. The foot is extended so that the firm pointed tip of the foot penetrates the substratum and, once fully extended, the epipodial collar is inflated to form a flange which anchors the foot firmly; muscular contraction of the foot muscles then pulls the shell into the sediment. These movements are probably responsible for the violent expulsion of water which removes faecal material from the posterior opening. The gradual extension of the foot when burrowing is presumed to aid the intake of water, the ciliary mechanisms alone not being thought to be sufficiently powerful to generate the observed inhalent current. Scaphopods are dioecious and fertilisation is external. **Eggs** are laid singly, developing through a **trochophore** larval stage to a **veliger** larva: after some 5–6 days, the veliger metamorphoses and becomes benthic in habit.

Little is known of scaphopod physiology. A **sub-radular organ** in the buccal cavity is thought to be chemoreceptive while the captaculae are important tactile organs. There is also a pair of **statocysts** located in the foot innervated from the cerebral ganglion not the pedal ganglion, despite lying near the latter.

Collection and Preservation

Scaphopods are usually collected in dredge or grab samples and, whenever possible, should be extracted from the sample in a living condition. Ideally animals should be narcotised in magnesium chloride solution or some other anaesthetising agent before fixation in 70% alcohol or 4% neutral formaldehyde solution. It should be noted that, once killed, the soft body is very likely to 'fall out' of the shell and if possible, individuals should be killed and stored in separate containers whenever mixed communities or shell and soft-body relationships are to be studied.

Classification

Order DENTALIOIDA Palmer, 1974
Family Dentaliidae Gray, 1874
Antalis entalis (Linnaeus, 1757)
Antalis vulgare (Da Costa, 1778)
Order SIPHONODENTALIOIDA Palmer, 1974
Family Siphonodentaliidae Simroth, 1894
Siphonodentalium lobatum (G. B. Sowerby, 1860)
Pulsellum lofotense (M. Sars, 1864)
Pulsellum affine (M. Sars, 1864)
Family Cadulidae Grant and Gale, 1931
Cadulus subfusiformis (M. Sars, 1865)
Cadulus jeffreysi (Monterosato, 1875)

Key

1. Aperture constricted .... *Cadulus* (**2**)
Aperture not constricted **3**

2. Shell slender; length to breadth ratio 5 to 6; dorsal (concave) side without or with only a very slight swelling near the middle .... *Cadulus subfusiformis* (p. 106)
Shell plump; length to breadth ratio about 4; dorsal (concave) side with a slight but distinct convex swelling near the middle .... *Cadulus jeffreysi* (p. 107)

3. Apex circular in section with a single, V-shaped notch; foot conical with an encircling epipodial sheath which is discontinuous, being slit on one side; shell apex may be finely sculptured by longitudinal striations .... *Antalis* (**4**)
Apex without single V-shaped notch; foot expanded distally into a symmetrical disc but without an epipodial sheath; shell surface without longitudinal striations **5**

4. Shell with fine longitudinal striations, particularly toward the apex, most visible in juvenile forms and often difficult to see in adults. Colour dull-white with black staining at the apex. Only slightly truncated .... *Antalis entalis* (p. 100)
Shell smooth, slightly more curved than *A. entalis* and posterior orifice bluntly truncated with, rarely, a short pipe. Apex usually reddish-brown .... *Antalis vulgare* (p. 100)

5. Apex of shell simple; pedal disc with long central filament .... (**6**)
Apex of shell cut into lobes or teeth; pedal disc without central filament .... *Siphonodentalium lobatum* (p. 103)

6. Length : breadth ratio about 12 or 13 .... *Pulsellum lofotense* (p. 104)
Length : breadth ratio about 6 or 7 .... *Pulsellum affine* (p. 104)

Order DENTALIOIDA Palmer, 1974

Scaphopods having shells, which may be large (up to some 150 mm long), with the maximum diameter at the anterior (oral) aperture. The shell surface may be smooth or strongly sculptured, usually with longitudinal ribs and/or transverse annulations. The radula bears a central tooth which is wider than high (about twice) and supports lateral teeth. Captacula are thin and numerous and the protractile foot has a distinct conical end and two, lateral, more or less subtrigonal, lobes.

Family DENTALIIDAE Gray, 1847

Shell medium to large and with a weak to pronounced curvature. The maximum diameter is at the anterior aperture. Sculpturing consists of 4–20 rather smooth, slightly sculptured ribs, with or without secondary ribs bearing striations or transverse lines. The apex is simple with a superficial notch on the concave side or with a deep fissure on the concave as well as the convex side. The shell is oval, circular or subpolygonal in section. The central radula-tooth has an entire cutting edge and is smooth and rather arched. The lateral tooth is strong and the marginal teeth are curved.

Genus ANTALIS H. and A. Adams, 1854

Shell small to medium-sized, circular or polygonal in section and weakly ribbed. Longitudinal riblets are frequently lacking in adult stages but present on the apical portion of the shell in juvenile stages. Apical orifice generally with a V-shaped notch on or near the concave side and generally bearing a solid plug with a central pipe or orifice, rarely simple.

Type species: *Antalis entalis* (L.).

Antalis entalis (Linnaeus, 1757)
(Fig. 25A–D)

Dentalium entalis Linnaeus, 1757
Dentalium entale Locard, 1898

A slightly curved, dull-white shell up to 40 mm long (Fig. 25A) and more or less circular in section. The apex is simple with a single, V-shaped notch (Fig. 25B) and the generally smooth surface may be finely ribbed near the apex only; this sculpturing may be strong in young animals. The apex of the shell is usually stained black and is never as abruptly truncated as in *A. vulgare* (Fig. 25E). A short but distinct pipe is often present (Fig. 25C). The foot is distinctly conical with two, laterally expanded, more or less subtrigonal lobes forming a split epipodial sheath (Fig. 25D). The radula bears a central tooth with an entire cutting edge and which is smooth and rather arched. The lateral teeth are strong and the marginal teeth are curved.

This species has been found in dead shell, gravel, sand, mud and clay deposits from depths of 6 m or more and is a North Atlantic species.

Antalis vulgare (Da Costa, 1778)
(Fig. 25E)

Dentalium vulgare Da Costa, 1778

Very similar to *A. entalis* but with a slightly more curved shell up to 60 mm long, with the posterior orifice distinctly truncated (Fig. 25E) and with a short pipe rarely visible. The apex is usually reddish-brown in colour. This is a more southerly species found from depths of 3 m or more.

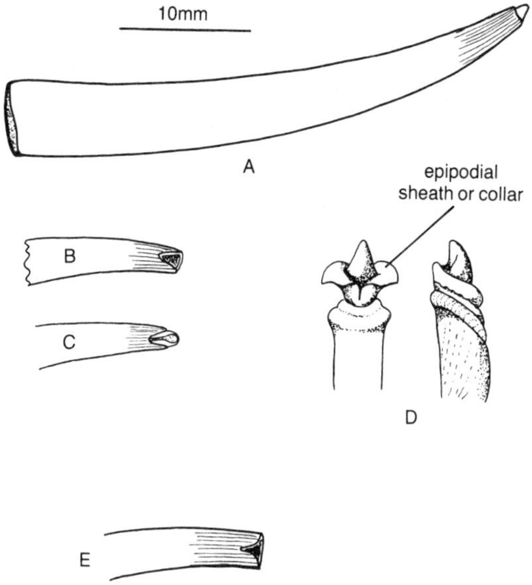

Fig. 25. *Antalis entalis*. A. Basic shell shape and distribution of sculpturing. B. Posterior of empty shell to show the notch. C. Posterior end of living specimen to show the pipe. D. The foot showing the epipodial sheath. E. *Antalis vulgaris*. Posterior of shell showing distinctly truncated end.

Order SIPHONODENTALIOIDA Palmer, 1974

Shell small to medium in size, usually less than 50 mm long, with the maximum diameter at or near the anterior (oral) end or midway along the length of the shell. Shell surface usually smooth or moderately sculptured with longitudinal striations and/or transverse annulations. The radula bears a central tooth which is higher than wide and smaller than the lateral teeth. Captacula are rather large, broad and less numerous than in the Dentalioida. The foot is reduced and formed by a retractile, filiform structure which often terminates in an expanded, circular pedal disc, with or without a central filament.

Family SIPHONODENTALIIDAE Simroth, 1894

Shell minute to moderate sized, commonly smooth and porcellaneous and rarely sculptured with longitudinal or annular markings. The maximum diameter of the shell is either at the anterior (oral) opening or very near to it. Two groups are distinct, those without a filament on the pedal disc (such as *Siphonodentalium*) and those with a central pedal filament (such as *Pulsellum*). Central radula tooth almost square.

Genus SIPHONODENTALIUM M. Sars, 1859

Shell moderately to strongly curved, slightly tapering and widest at the anterior aperture. Shell circular in section, rarely compressed dorso-ventrally: surface smooth except for growth lines. Apex large, cut into lobes. Foot has a distinct pedal disc, concave medially and lacking a central filament or process.

Type species: *Siphonodentalium lobatum* (G. B. Sowerby).

Siphonodentalium lobatum (G. B. Sowerby, 1860) (Fig. 26A, B)

Dentalium lobatum G. B. Sowerby, 1860

A moderately curved shell up to 20 mm long (Fig. 26A) and circular in section with the apex distinctly notched into lobes or teeth. The shell surface has a smooth, shiny, porcelaneous appearance. The foot is expanded distally into a symmetrical disc but lacks any central filament (Fig. 26B). The central tooth of the radula is complex and variable while the lateral teeth are curved and have three distinct cusps: the marginal teeth are straight with a longitudinal carination.

They are found from depths of 30 m or more and have only been rarely recorded from British waters.

Genus PULSELLUM Stoliczka, 1868

Shell moderately to strongly curved, slightly tapering and with the largest diameter at the oral aperture. Shell usually circular in section, rarely compressed dorso-ventrally. Surface smooth with only growth lines. Apex simple and without lobes or slits. Pedal disc convex medially and with a distinct central filament.

Type species: *Pulsellum lofotense* (M. Sars).

Pulsellum lofotense (M. Sars, 1864)
(Fig. 26C)

Siphonodentalium lofotense M. Sars, 1864

A small (up to 6 mm long) shell of moderate curvature and circular section (Fig. 26C): the length : breadth ratio is 12–13 : 1. The apex is simple and sculpturing is variable, mainly annular in character. The pedal disc bears a distinct central filament. The central tooth of the radula is simple while the marginal teeth have a lateral carination.

This is a deep water form from depths of 200 m or more.

Pulsellum affine (M. Sars, 1864)
(Fig. 26D)

Siphonodentalium affine M. Sars, 1865

As *P. lofotense* except that the shell is shorter (up to about 4 mm long) and relatively broader (Fig. 26D) with a length : breadth ratio of about 6–7 : 1. A deep water form only.

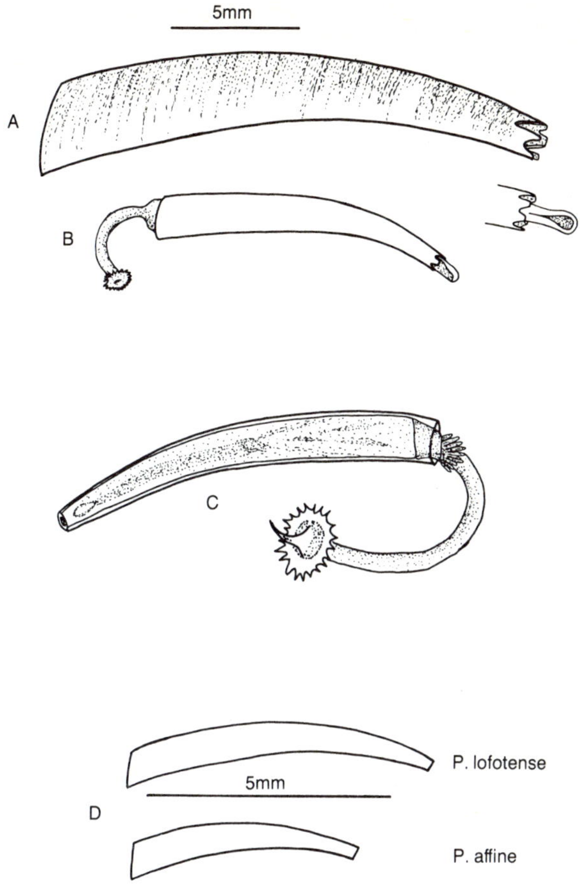

Fig. 26. *Siphonodentalium lobatum*. A. Shell showing lobes at posterior end. The inset shows the pipe present in living specimens. B. Entire animal showing the pedal disc. C. *Pulsellum lofotense*. Entire specimen showing the pedal disc. D. Comparison of shell profile and dimensions of *P. lofotense* and *P. affine*.

Family CADULIDAE Grant and Gale, 1931

Small to medium sized animals with a weak to moderate curvature and rounded or dorso-ventrally compressed in section. Aperture constricted so that the maximum diameter is towards the middle of the shell. The shell is smooth, shining and with little sculpturing. The foot is vermiform but does not bear a central filament.

Genus CADULUS Philippi, 1844

Shell small to moderate, arcuate and with the maximum diameter near the middle or between the median portion and the oral aperture. The aperture generally constricted and never the region of maximum diameter. Shell surface smooth, rarely sculptured with longitudinal striae or minute annular rings. Cross-section circular or oval, never angular. Apical orifice simple or with 2–4 notches, orifice often constricted with a ledge within the opening. Shell texture vitreous and transparent/translucent. Foot vermiform with a pedal disc but no filament.

Type species: *Cadulus ovulum* Philippi.

Cadulus subfusiformis (M. Sars, 1865)
(Fig. 27A, B)

Siphonodentalium subfusiforme M. Sars, 1865.

A very small shell only some 3 mm long and constricted at both ends (Fig. 27A). The maximum diameter is more or less in the middle of the shell and the dorsal (concave) side of the shell has only a very slight median swelling (Fig. 27B). The length : breadth ratio is about 5–6 : 1. The shell is smooth and shiny and has longitudinal striations/annulations in the posterior region. The apex is simple. The pedal disc lacks any central filament. The radula bears a simple central tooth which has thickened lateral margins and a rather straight, thin, denticulated superior margin. The lateral teeth bear three distinct cusps and a number of smaller denticulations. The marginal teeth have a pronounced antero-lateral carination.

This species is found at depths of 80 m or more.

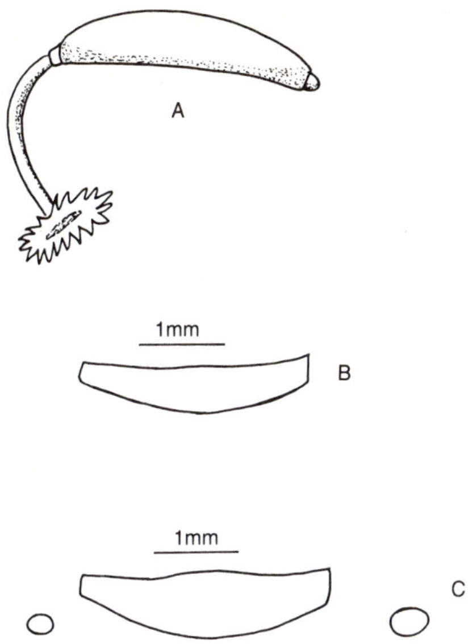

Fig. 27. A. Entire specimen of *Cadulus subfusiformis* to show pedal disc. B. Profile of shell of *C. subfusiformis*. C. Profile and cross sections of shell of *C. jeffreysi*.

Cadulus jeffreysi (Monterosato, 1875)
(Fig. 27C)

Helonyx Jeffreysi Monterosato, 1875
Cadulus Jeffreysi Locard, 1898

A small shell up to 3.5 mm long and with a length : breadth ratio of about 4 : 1. The shell is more plump than *C. subfusiformis* and the dorsal (concave) surface has a slight but distinct median swelling (Fig. 27C).

Found only from deeper waters than *C. subfusiformis*, from about 250 m or more.

Glossary of Special Terms

Acicular Needle-shaped; slender and sharp-pointed.

Articulamentum The inner, usually hard and semi-porcelaneous, shell layer of chiton valves, generally projecting past the tegmentum on the sides and front of the valves to form insertion plates and the sutural lamina. Bounded internally by the hypostracum.

Basal Plate The medioventral supporting element of the caudofoveate radula; it may or may not be partially sclerotised (hardened).

Cloaca Terminal or sub-terminal cavity into which open both the excretory and reproductive ducts.

Ctenidia (sing. *ctenidium*) The typically bi-pectinate molluscan gill formed of a central filament with two rows of leaf-like branches.

Digestive Gland The large, usually median, organ opening from the molluscan stomach, which carries out the principal digestion processes; this is an organ of mainly intracellular digestion. (See also Midgut gland).

Dioecious Unisexual, the male and female reproductive organs being borne on separate individuals.

Distich/Distichous A radula in which the teeth are arranged in two rows.

Divaricate Diverging rows.

Epipodial sheath The membranous, often expanding, collar found in dentaloid scaphopods which is used in burrowing.

Epizoic Dwelling on another animal.

Foot The single, median structure used for creeping or burrowing in molluscs. It varies markedly in significance between and within classes.

Foot Shield The area adjacent to the mouth of caudofoveates which remains uncovered by the mantle. It comprises a cuticularised ciliar epithelium which may be post-oral, pre-oral or lateral in position.

Girdle Flexible muscular integument of chitons, variously ornamented, which forms a 'skirt' allowing the animal to closely oppose itself to irregular surfaces.

Head Valve The anterior shell valve of chitons, usually numbered I.

Holobranch The condition in which the ctenidia extend the whole length of the pallial groove.

Hypodermis Layer of cells which may secrete some strengthening elements lying immediately below the epidermis in some solenogastres.

Hypostracum The thin, innermost calcareous layer of a chiton shell valve.

Imbricate Overlapping, as in tiles on a roof.

Insertion Plate Narrow marginal extension of the articulamentum of chitons in the head and tail valves and at the sides of intermediate valves. They project into the girdle and may also be called **laminae of insertion**.

Intermediate valves Any one of the shell valves of a chiton between the head and tail valves. Numbered, in order, II to VII, from the anterior end.

Jugal Area (or tract) Upper surface of a chiton shell valve immediately adjacent to the jugum, sculptured differently from the rest of the surface in some forms. Also called the **dorsal area**.

Jugal Sinus The depression or 'bay' between the sutural laminae in chitons.

Jugum Longitudinal ridge in the middle of an intermediate shell valve in chitons.

Lateral Triangle (or area) Portion of the upper surface of an intermediate valve in chitons, commonly triangular in shape, lying at the side and towards its posterior, usually sculptured diagonally and set off from the remainder of the upper surface by a diagonal ridge of varying prominence.

Mantle Cavity A typically posterior cavity formed from flaps of the mantle proper, and which usually contains the openings of the alimentary, excretory and reproductive systems as well as the ctenidia (gills).

Median Triangle (or area) Portion of upper surface of an intermediate valve in chitons, commonly triangular in shape, lying at the side and towards its anterior, usually sculptured antero-posteriorly.

Megalaesthete Large cellular organ found in large numbers in the tegmentum of chiton shell valves and terminating dorsally in apical caps.

Merobranch The condition in which the ctenidia extend only part of the length of the pallial groove, being absent from the anterior region.

Midgut Gland The caudofoveate equivalent of the molluscan digestive gland; it is a large medioventral organ primarily of intracellular digestion.

Mucro Point or projection on the chiton tail valve, usually marking a separation between the configuration of the central and posterior areas. The mucro may be prominent or obsolete and variable in size, shape and position.

Notch Ray (also called **slit ray**) A shallow groove or row of pores or pits in chitons, extending from a notch (slit) to the apex of the valve on the ventral surface.

Pallial Cavity The cavity between the mantle and the foot in most molluscs, often containing ctenidia and sensory organs. In the minor mollusc classes, it is often used to describe the posterior cavity which is also a cloaca.

Pedal Disc The distal expansion of the vermiform foot of some scaphopods; usually circular and sometimes with a median filament.

Pedal Groove The ventral groove in solenogastres which is variable in structure and which may be equivalent to the pallial groove in other molluscan classes.

Pedal Pit A sensory depression, usually containing ciliated cells, at the anterior end of the pedal groove in solenogastres.

Periostracum The outermost layer of a chiton shell valve, formed of a proteinaceous material and probably protective in function.

Pharynx The muscular area between the buccal cavity and the oesophagus.

Pipe The siphon-like organic extension which protrudes from the posterior opening in some scaphopods.

Radula The bilaterally symmetrical, tongue-like ribbon containing chitinous denticles used for feeding in some molluscs. It is a rasping structure capable of being protruded from the mouth when grazing.

Shell Valve (or Plate) A discrete shell unit of chitons usually numbered I to VIII from the anterior end.

Spatulate Paddle-shaped.

Spicula/Spicule Calcareous spine-like structures embedded in the mantle of some molluscan classes. They vary widely in size, shape and abundance and may occur together with scales.

Squamate/Squamiform Scale-shaped in form.

Sutural lamina Sharp, plate-like anterior projections of the articulamentum of chitons, varying in prominence and extending from both sides of the anterior edge of intermediate and tail valves to articulate with the valve in front. They may be separated by the jugal sinus or joined by a laminar extension of the articulamentum.

Symphyse A ligamental joining of two or more parts.

Tail Valve The posterior valve of a chiton, numbered VIII in the series.

Tegmentum The outer, usually softer and somewhat more porous, calcareous layer of a chiton shell valve just below the periostracum; this layer contains the majority of the aesthete system.

Trochophore Ciliated, usually planktonic, larval stage of molluscs and several other invertebrate phyla.

Uncinate Blade The large scraping tooth of the radula; in chitons this is hardened by deposition of a layer of magnetite.

Veliger Planktonic larval stage which develops from a molluscan trochophore but in which the foot, mantle, shell and other adult organs are present.

Acknowledgements

The authors would like to express their gratitude to the malacological staff at the British Museum and the Royal Scottish Museum for access to their collections of Molluscs.

We also wish to thank Dr J. P. Hartley for specimens of Solenogastres and Caudofoveates loaned to us during the preparation of this *Synopsis* and to the Conchological Society of Great Britain and Ireland for their provision of copies of the distribution maps for relevant species.

References

Caudofoveata and Solenogastres

Gage, R. 1972. Community structure of the benthos in Scottish sea-lochs. *Mar. Biol.*, **14,** 281–297.

Hansen, G. A. 1889. Neomenia, Proneomenia und Chätoderma. *Bergens Mus. Aarsberetn,* 1888.

Hoffman, S. 1949. Studien uber das Integument der Solenogastren. *Zool. Bidr Uppsala*, **27,** 293–427.

Hyman, L. H. 1967. *The Invertebrates IV. Mollusca I. Aplacophora, Polyplacophora, Monoplacophora, Gastropoda.* 792 pp. McGraw-Hill Book Co., New York.

Ivanov, D. L. 1981. *Caudofoveatus tetradens* as a generic and specific term, and the diagnosis of taxa in the sub-class Caudofoveata (Mollusca, Aplacophora). *Zool. Journal*, **LX,** 18–28.

Kowalevsky, A.-O. and Marion, A. F. 1887. Contributions a l'histoire des Solenogastres ou Aplacophores. *Ann. Mus. Hist. nat. Marseille*, **3,** 1–76.

Muus, B. J. 1959. Skallus, Søtaender, Blaeksprutter. *Danm. Fauna*, **65,** 1–239.

Salvini-Plawen, L. von 1969. Solenogastres und Caudofoveata (Mollusca, Aculifera): Organization und phylogenetische Bedeutung. *Malacologia*, **9,** 191–216.

Salvini-Plawen, L. von 1971. Schild- und Furchenfusser (Caudofoveata und Solenogastres). *Die Neue Brehm-Bucherei*, **441,** 1–95.

Salvini-Plawen, L. von 1975. Mollusca; Caudofoveata. *Marine Invertebrates of Scandinavia*, **4,** 1–55.

Salvini-Plawen, L. von 1978. Antarktische und subantarktische Solenogastres-Eine monographie: 1898–1974. *Zoologica, Stuttgart*, **128,** 1–315.

Scheltema, A. H. 1978. Position of the class Aplacophora in the phylum Mollusca. *Malacologia*, **17,** 99–109.

Polyplacophora

Baxter, J. M. 1982. Allometric and morphological variations of whole animal and valve dimensions of the chiton *Lepidochitona cinereus* (L) (Mollusca, Polyplacophora). *J. Moll. Stud.*, **48,** 275–282.

Baxter, J. M. and Jones, A. M. 1984. The valve morphology of *Callochiton achatinus* (Mollusca: Polyplacophora). *J. Zool. London*, **202,** 549–560.

Baxter, J. M. and Jones, A. M. 1986. Allometric and morphological characteristics of *Tonicella marmorea* (Fabricus, 1780) populations (Mollusca: Polyplacophora: Ischnochitonidae). *J. Linn. Soc. (Zool)*, (in Press).

Bergenhayn, J. R. M. 1955. Die fossilen schwedischen Loricaten nebst einer vorlaufigen Revision des systems der gauzen Klass Loricata. *Lunds Univ. Arssk.*, **51,** 1–41.

Kaas, P. 1981. Scandinavian species of *Leptochiton* Gray, 1847 (Mollusca, Polyplacophora). *Sarsia*, **66,** 217–229.

Kaas, P. 1985. The genus *Acanthochitona* Gray, 1821 (Mollusca, Polyplacophora) in the north-eastern Atlantic Ocean and in the Mediterranean Sea, with designation of neotypes of *A. fascicularis* (L. 1767) and of *A. crinita* (Pennant, 1777). *Bull. Mus. natn. Hist. nat., Paris* 4ser, **7A,** 579–609.

Linnaeus, C. von 1758. *Systema naturae per regna tria naturae, secundem classes, ordines, genera, species cum characteribus, differentis, synonymis, locis.* Ed. 10, 1, 823 pp.

Matthews, G. 1953. A key for use in the identification of British chitons. *Proc. Malac. Soc. Lond.*, **29,** 241–248.

Pilsbry, H. 1892. *Monograph of the Polyplacophora.* In Manual of Conchology 14 (Ed. G. Tryon Jnr.). 350 pp. Philadelphia.

Thiele, J. 1929. *Classis Loricata.* In Handbuch der systematischen Weichtierkunde, **1,** 1–12.

Scaphopoda

Bilyard, G. R. 1974. The feeding habits and ecology of *Dentalium entale stimpsoni* Henderson (Mollusca: Scaphopoda). *The Veliger*, **17,** 126–138.

Dinamani, P. 1954. Feeding in *Dentalium conspicuum. Proc. Malac. Soc. Lond.*, **36,** 1–5.

Emerson, W. K. 1962. A classification of the scaphopod molluscs. *J. Palaeont.*, **36,** 461–482.

Morton, J. F. 1959. The habits and feeding of *Dentalium entalis. J. mar. biol. Ass. U.K.*, **38,** 225–238.

Palmer, C. P. 1974. A supraspecific classification of the Scaphopod Mollusca. *The Veliger*, **17,** 115–123.

Palmer, C. P. 1979. *An Introduction to the Scaphopoda.* Conch. Soc. of Great Britain and Ireland; Papers for students No. 17, 5 pp.

Yonge, C. M. 1937. Circulation of water in the mantle cavity of *Dentalium entalis. Proc. Malac. Soc. Lond.*, **22,** 333–337.

Appendix

Distribution maps for the species described in this *Synopsis* (Reprinted from 'The Sea Area Atlas of the Marine Molluscs of Britain and Ireland' published by the Nature Conservancy Council/Conchological Society).

(● – recorded live, post-1950; ◑ – recorded live, pre-1951; ○ – recorded as shell only, any date).

CAUDOFOVEATA and SOLENOGASTRES

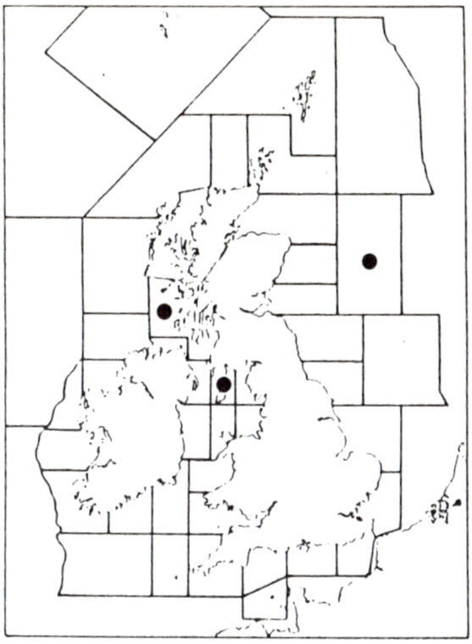

SCUTOPUS VENTROLINEATUS

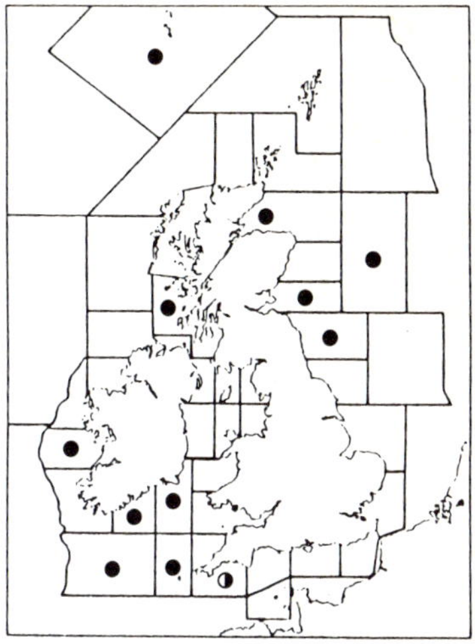

CHAETODERMA NITIDULUM

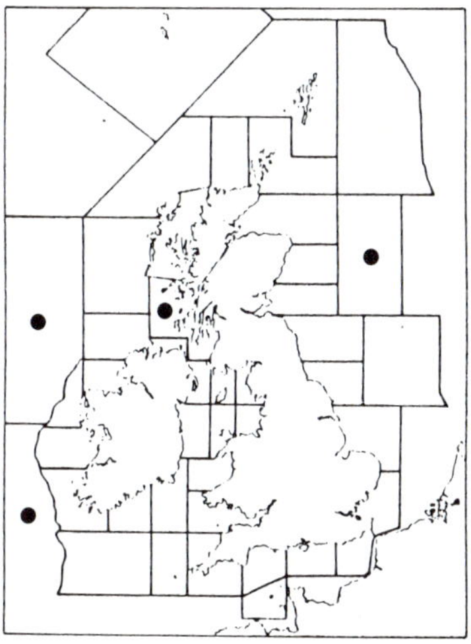

FALCIDENS CROSSOTUS

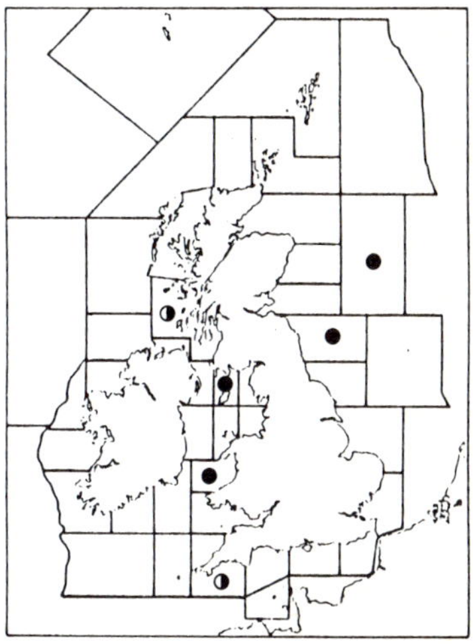

NEMATOMENIA BANYULENSIS

SOLENGASTRES

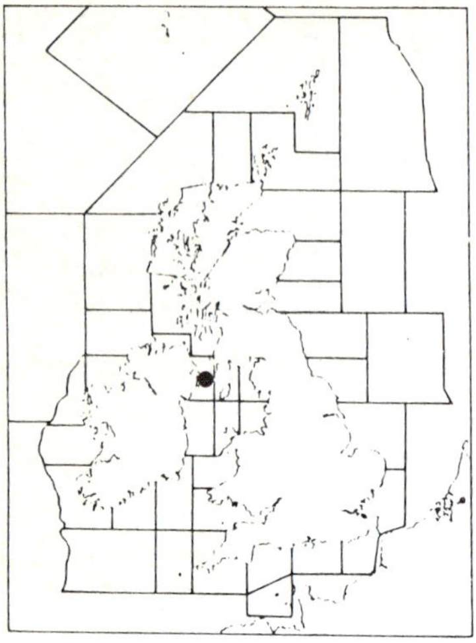

LEPIDOMENIA HYSTRIX

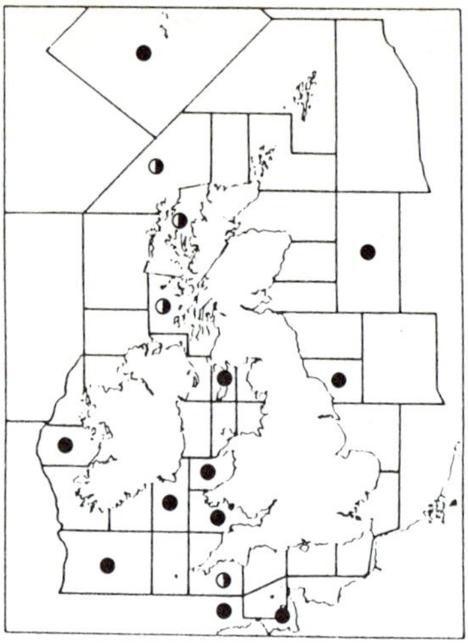

NEOMENIA CARINATA

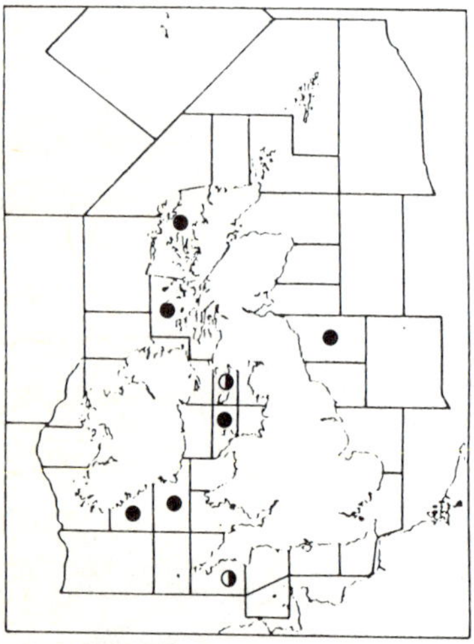

RHOPALOMENIA AGLAOPHENIAE

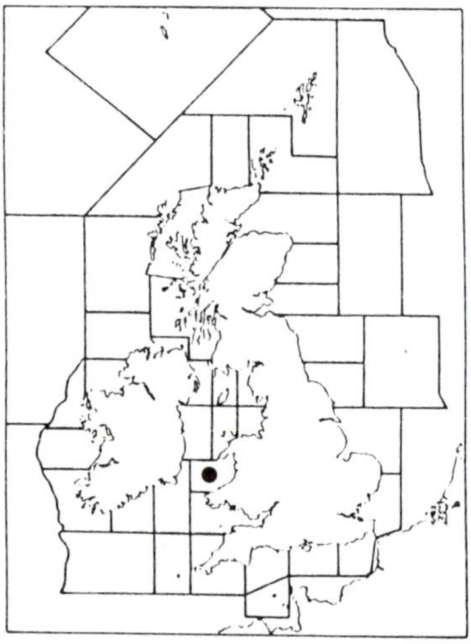

ELEUTHEROMENIA SIERRA

POLYPLACOPHORA

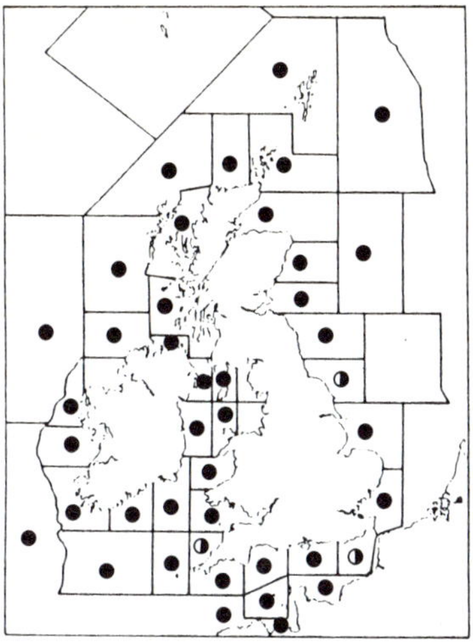

LEPTOCHITON ASELLUS

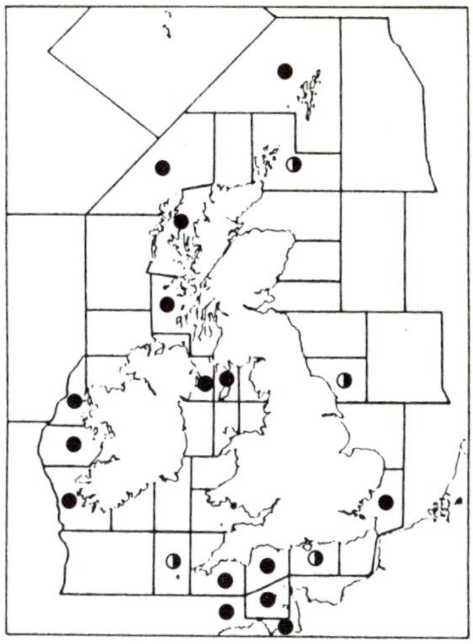

LEPTOCHITON CANCELLATUS

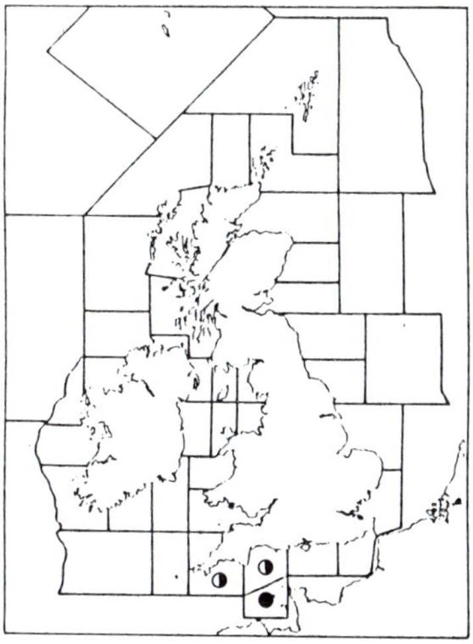

LEPTOCHITON SCABRIDUS

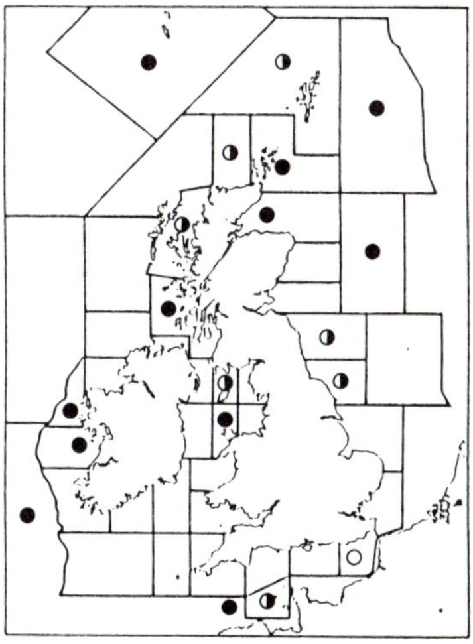

HANLEYA HANLEYI

POLYPLACOPHORA

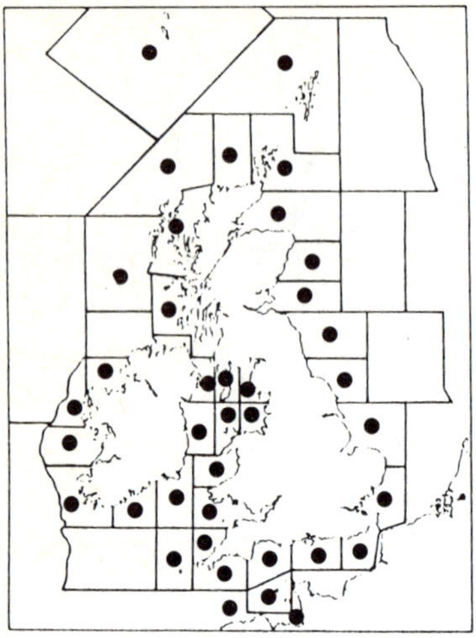

LEPIDOCHITONA CINEREUS

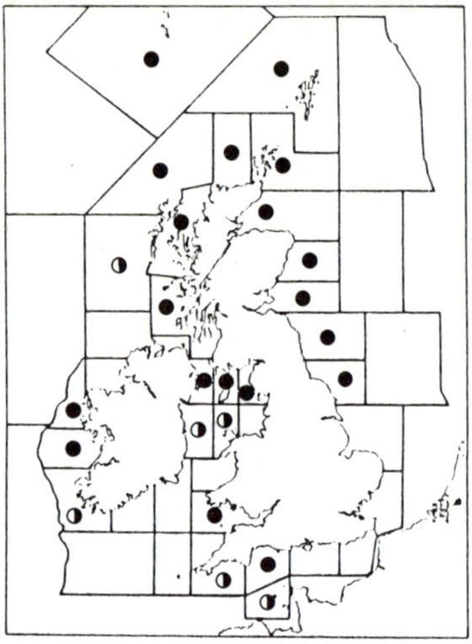

TONICELLA RUBRA

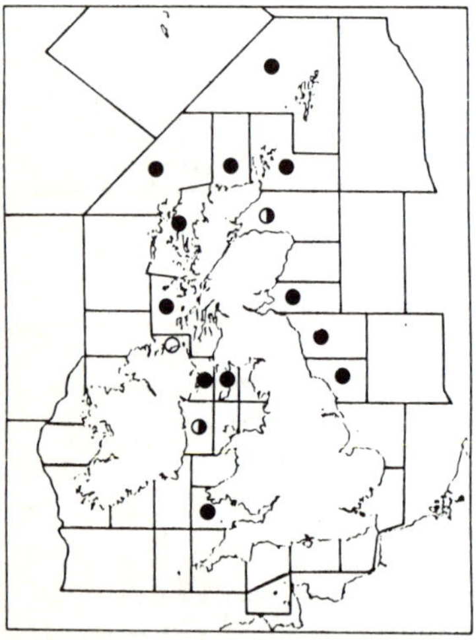

TONICELLA MARMOREA

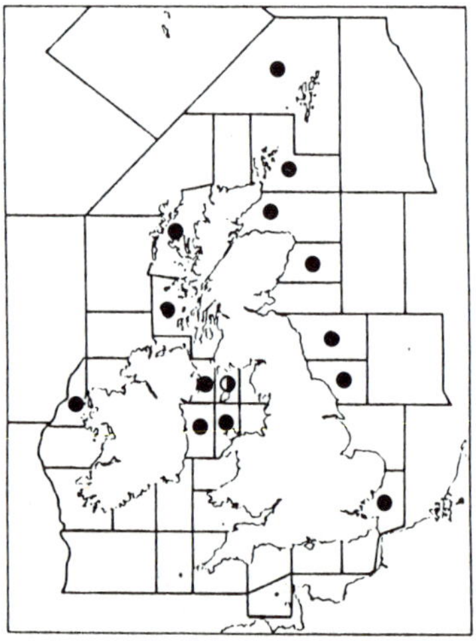

ISCHNOCHITON ALBUS

POLYPLACOPHORA

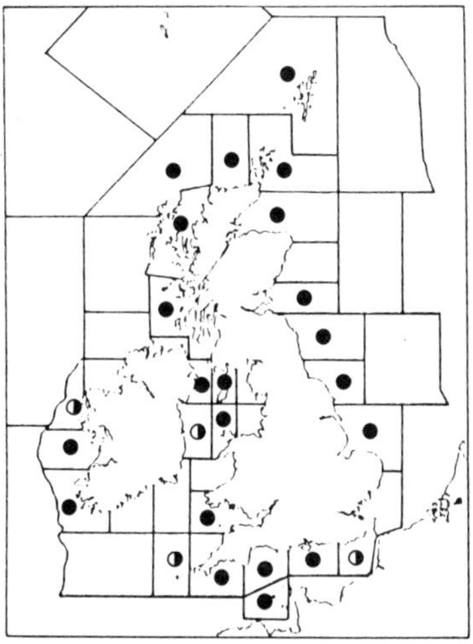

CALLOCHITON ACHATINUS

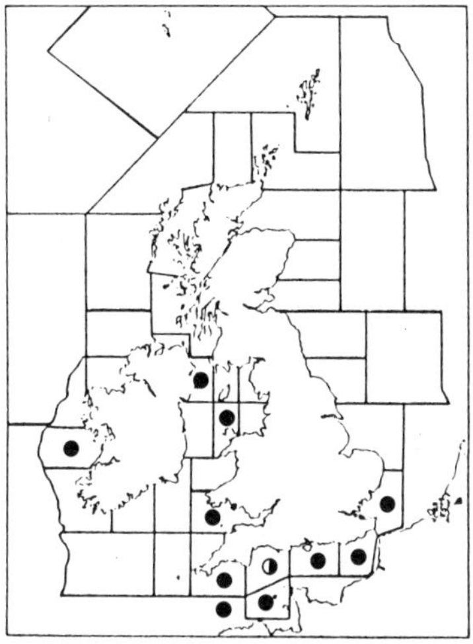

ACANTHOCHITONA FASCICULARIS

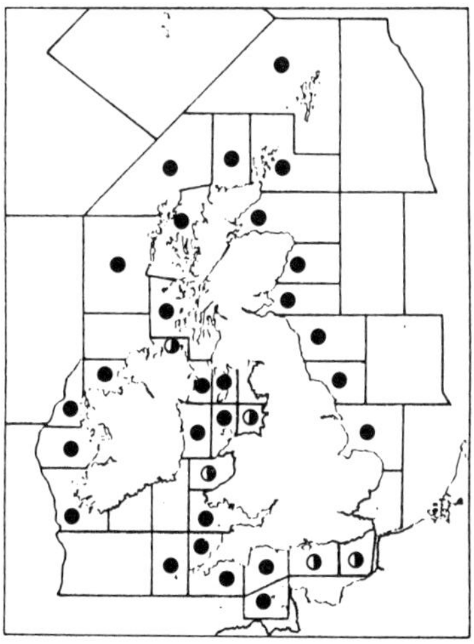

ACANTHOCHITONA CRINITUS

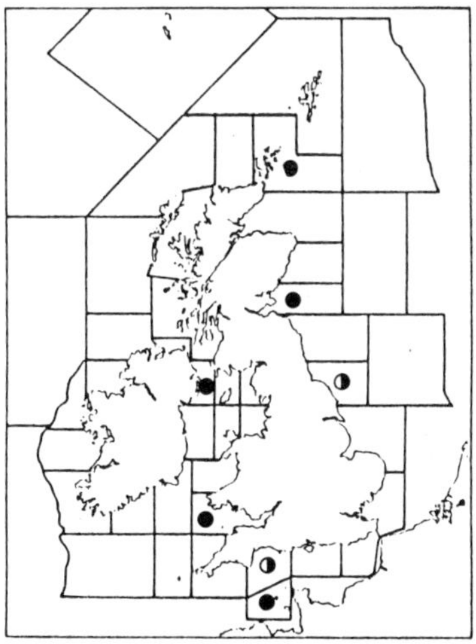

ACANTHOCHITONA DISCREPANS

SCAPHOPODA

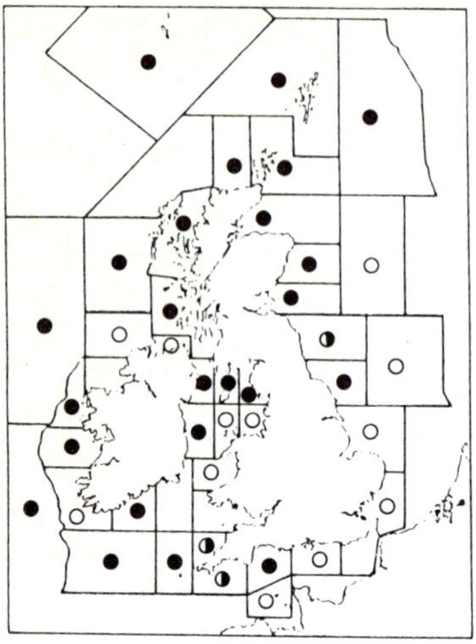

ANTALIS ENTALIS

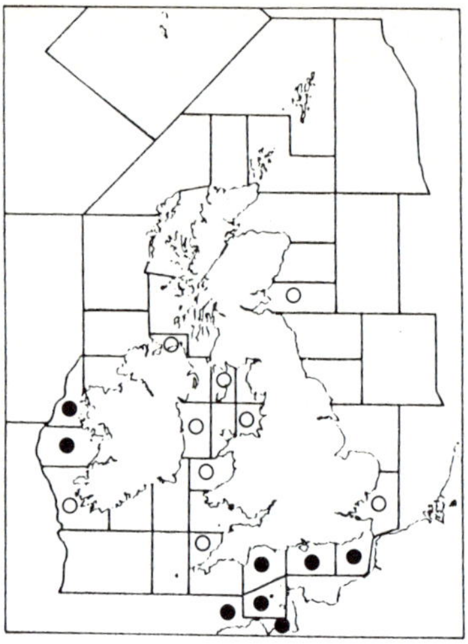

ANTALIS VULGARE

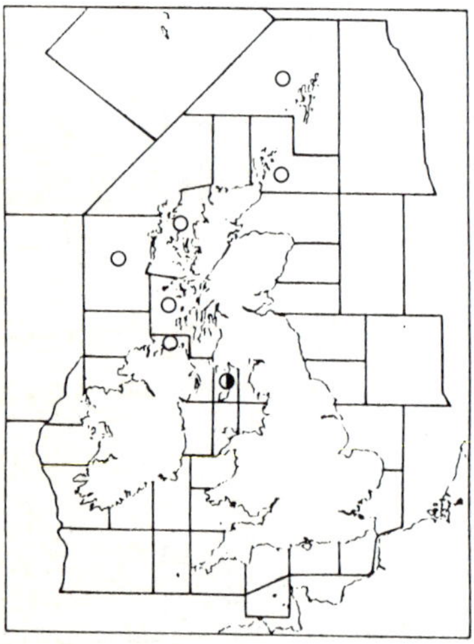

PULSELLUM LOFOTENSE

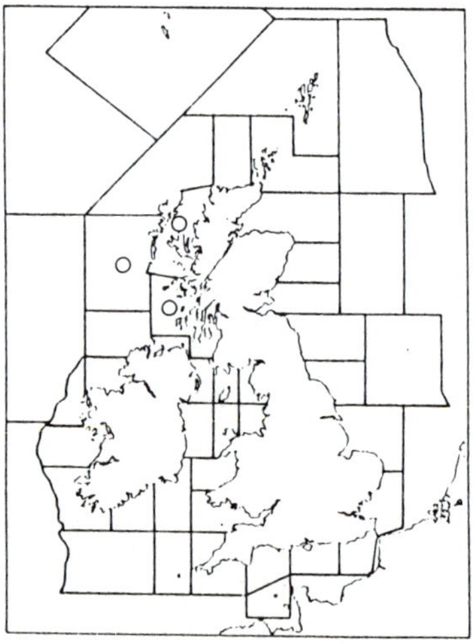

PULSELLUM AFFINE

SCAPHOPODA

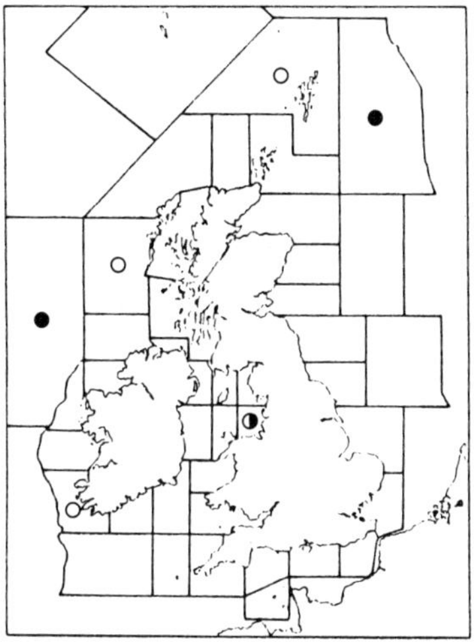

CADULUS JEFFREYSI

Index of Families, Genera and Species

For species and genera the correct names are in *italics*; synonyms are in roman.

CAUDOFOVEATA
CHAETODERMATIDAE 14
Chaetoderma 14
nitidulum 14
Crystallophrisson nitens 14
nitidulum 14
Falcidens 16
crossotus 16
LIMIFOSSORIDAE 10
Scutopus 10
ventrolineatus 10

SOLENOGASTRES
DONDERSIIDAE 26
Dondersia banyulensis 26
Eleutheromenia 36
sierra 36
LEPIDOMENIIDAE 28
Lepidomenia 28
hystrix 28
Myzonemia banyulensis 26
Nematomenia 26
banyulensis 26
NEOMENIIDAE 30
Neomenia 30
carinata 30
dalyelli 32
Paramenia sierra 36
PARARHOPALIIDAE 36
Proneomenia aglaopheniae 34
RHOPALOMENIIDAE 33
Rhopalomenia 34
aglaopheniae 34
Solenopus dalyelli 32
Solenopus nitidulus 30

POLYPLACOPHORA
Acanthochites adansoni 88
bouvieri 88
carinatus 84
communis 84
hamatus 84
Acanthochiton discrepans 84
communis 84
fascicularis 88
oblongus 88
Acanthochitona 82
bonairensis 84
fascicularis 82, 84
communis 84
crinitus 82, 88
discrepans 83, 90
ACANTHOCHITINIDAE 81
Boreochiton ruber 66
marmoreus 70
Callochiton 78
achatinus 78
doriae 78
laevis 78
Chiton achatinus 78
aselloides 74
asellus 50
cancellatus 54
cimicinus 50
cinereus 50, 62
crinitus 84, 88
discors 78
discrepans 84
doriae 78
euplaea 78
fascicularis 84, 88
fascicularis *var* major 84
fascicularis *var* minor 88
fascicularis *var* attenuata 88
fuscatus 50
hanleyi 58
laevis 78
marginatus 62
marmoreus 70
minimus 50, 74
onyx 88
oryza 74
ruber 66
sagrinatus 74
scabridus 56

Chiton contd.
scoticus 50
septemvalvis 78
tuberculatus 54
Craspedochilus marginatus 62
HANLEYIDAE 58
Hanleya 58
hanleyi 58
Hanleya debilis 58
dalli 58
ISCHNOCHITONIDAE 62
Ischnochiton 74
albus 74
marginatus 62
ruber 66
Laminoplax dalli 58
Lepidochiton cinereus 62
marmorea 70
ruber 66
Lepidochitona 62
cinereus 62
cinerea 62
marmorea 70
LEPIDOPLEURIDAE 49
Lepidopleurus asellus 50
cancellatus 54
carinatus 58
scabridus 56
Leptochiton 49
asellus 50
cancellatus 54
scabridus 56
ruber 66
Lophyrus albus 74
Tonicella 66
ruber index 66
rubra 66
marmorea 70
submarmorea 70
Tonicia rubra 66
marmorea 70
Trachydermon cinereus 62
ruber 66

SCAPHOPODA
DENTALIIDAE 98
Antalis 99
entalis 100
vulgare 100
CADULIDAE 106
Cadulus 106
jeffreysi 107
Jeffreysi 107
subfusiformis 106
Dentalium entalis 100
entale 100
lobatum 103
vulgare 100
Helonyx Jeffreysi 107
Pulsellum 104
affine 104
lofotense 104
SIPHONODENTALIIDAE 102
Siphonodentalium 102
affine 104
lobatum 103
lofotense 104
subfusiforme 106